Constructing "The Beginning"

Discourses of Creation Science

Constructing "The Beginning"

Discourses of Creation Science

Simon Locke

Kingston University, United Kingdom

Routledge
Taylor & Francis Group

NEW YORK AND LONDON

First published 1999 by Lawrence Erlbaum Associates, Inc.

Published 2010 by Routledge
605 Third Avenue, New York, NY 10017
4 Park Square, Milton Park, Abingdon, Oxon OX14 4RN

*Routledge is an imprint of the Taylor & Francis Group,
an informa business*

Copyright © 1999 by Taylor & Francis

Cover design by Kathryn Houghtaling Lacey

Library of Congress Cataloging-in-Publication Data
Locke, Simon.
Constructing "the beginning" : discourses of creation science / by
    Simon Locke.
        p. cm. — (LEA's communication series)
Includes bibliographical references and index.
ISBN 0–8058–2346–8 (alk. paper). — ISBN 0–8058–2347–6
    (alk. paper : pbk.)
1. Creationism. 2. Science—Philosophy. I . Title. II. Series.
BS651.L63   1998
231.7'652—dc21
                                                    98–13598
                                                       CIP

ISBN 13: 978-0-8058-2347-9 (pbk)

For Myradh

Contents

Contents

Preface

Creation science is the target of much attack these days, from both within and without of the orthodox scientific community. In this book, however, I have tried to take a different approach. This is not an attack on creationism; but this should not be taken to mean it is a defense, either. In some respects, my interest is not really in creationism at all; rather, it is in questions of the role and significance of science in modernity, or, broadly, the public understanding of science. My approach to this issue is a discursive and rhetorical one. Creationism is treated as a case study of the argumentative engagement between science and nonscience (an argument that incorporates disputes over this boundary), which, in my view, is as central to the commonsense lifeworld of modernity, as it is to the lives of its intellectuals. An important dimension of the public meaning of science in modernity is its limits and its relations with other modes of thought and belief, which continue to survive as discourses in the wider culture. Creationism is merely one example of this general feature.

The book is made somewhat like a sandwich. The inner layers consist of a detailed empirical analysis of the discourse of creation scientists, as represented in pamphlets published by the British Creation Science Movement. This analysis is enclosed in two layers of theoretical discussion, addressing issues in the public understanding of science and, more broadly, the representation in sociological theory of the role and position of science in relation to modern society and culture.

There are various ways of eating a sandwich, and there are various ways of reading this book. Chapter 1—the top slice—sets out the theoretical and methodological background. It discusses current issues in the public understanding of science in relation to traditional sociological views of the impact of science on modernity, through rationalization, and the contrasting view derived from the sociology of scientific knowledge, which points to the likelihood of a much more complex and variable relationship than rationalization proposes.

I use this contrast to highlight what I call the *dilemmatic of science* in modernity. This refers to the tension between the universal knowledge claims made by or on behalf of science, and the fact that science arises from a particular society (modernity) and is the result of particular actions taken by particular people (scientists) at particular times in particular places. This tension is played out—as we well know—in current intellectual debates; but it is also played out in wider public contexts.

This forms my basic approach to creation science. In addition, I utilize a version of discourse analysis, which draws heavily on the work of Jonathan Potter and his various collaborators, in combination with the rhetorical-dilemmatic approach of Michael Billig and his collaborators responsible for the book *Ideological dilemmas* (1988). I use this to set out a background for the analysis of the discourse and rhetoric of creation scientists in relation to the discourse of orthodox scientists and sociologists seeking to advance explanatory accounts of the apparent resurgence of creationism in recent decades.

As such, my approach has much in common with the recent "rhetoric of inquiry" literature, especially the insightful analyses of Lawrence Prelli and Charles Alan Taylor. However, it is oriented to a different set of questions. My concern is less with the rhetoric of science per se, and not even so much with the rhetoric of demarcation, although this does occupy a substantial position in the analysis (especially chapter 3). My interest is rather more in the public rhetorics that surround science and draw upon it as a resource. Too often, these are accepted as conforming to the simple monolithic representations argued for by generations of academics—especially, but by no means solely, many publicly prominent scientists. Indeed, this is sometimes assumed also by rhetoricians whom, one would have thought, ought to, if not know, at least suspect different. This study, then, is intended to begin to consider what a rhetorically informed perspective can offer in opening up for more careful consideration the complexities of public argumentation about and around science. As such, it makes something of a counterpoint to Prelli and Taylor—or, to continue the culinary metaphor, a sandwich from a different, sociologically informed, kitchen.

The content of this sandwich is the analysis of creationist rhetoric, which begins in chapter 2 and continues through to chapter 5. If you prefer to eat your sandwiches inside out, I recommend you start with chapter 2. You should be able to follow the argument and the detailed analysis that follows without too much trouble. The analysis itself focuses on three main points. Creationists, I argue, are faced with three problems of discursive management: First, there is the problem of a competing account of reality (the world), in the form of evolution. Second, there is the problem of competing accounts of the Bible (the Word), in the form of different versions of Christianity, especially those that seek to make the biblical text compatible with evolution. Third, both of these problems must be managed together, in such a way that creationists' own version(s) of the world and of the Word are compatible—a compatibility achieved through what I call a *discursive syncretism.*

If, on the other hand, you like to eat all the bread in your sandwiches before you start on the innards, after chapter 1, you can go straight to chapter 6. This is also concerned with the dilemma of science. It focuses on three recent theoretical accounts of modernity that attempt to accommodate the apparent fragmentation of contemporary culture and its reluctance to fit easily into the framework of rationalization. Although far from exhaustive, these accounts, of Habermas, Bell, and Lyotard, cover the broad range of approaches from each of critical theory, postindustrialism, and postmodernism, respectively. I show that each approach attempts to cope with the exhaustion of rationalization, the failure of its monolithic representation of science, but is unable to do so without radical modification; modification which points toward a view of science as variable and multiple—like that shown by sociology of scientific knowledge (SSK), in fact. However, although the nettle is seen, it is not grasped. These theories play out the dilemma of science, but do not resolve it. Thus, they remain trapped within essentially the same discourse of science as the creationists; as such they are unable to reflect adequately upon this discourse and comprehend its dilemmatic nature.

The final chapter brings together the strands of the argument to further develop the implications of the dilemma of science for the public understanding of science, through the idea of *science as a cultural resource* and its possible relation to other such cultural resources within modernity—such as Christianity. It is suggested that much so-called antiscience could be made sense of in these terms and proposes further research in this direction. This particular layer of the sandwich may be eaten at any point after the others.

If you finish the whole thing (or maybe before you start it!), you might like to know that it began life as a PhD thesis—itself a sandwich, but from still another kitchen. Some ingredients have been removed and some new ones added. Whether this makes for an altogether different product or simply a variation on a continuing theme is an open question.

Over the years of its manufacture, a lot of debt to a lot of people has accrued, far too much to mention all here. Particular gratitude is owed to my two supervisors, Clive Ashworth and Derek Layder. Each helped in rather different ways, but this study would never have happened without them. Derek, I must thank in particular for the suggestion to apply my somewhat inchoate effort at theoretical synthesis to the particular case of creationism and for his continuing support and encouragement despite our divergent perspectives.

Where Derek left off, I have been especially fortunate in having Jonathan Potter take over. I am immensely in his debt, in numerous ways. I hope he will find that the resulting product is worthy of the efforts he has made on my behalf.

Thanks are due also to my colleagues at Kingston University for their support, not least to the Faculty of Human Sciences for allowing me the time to write. I have also had valuable discussions which have helped to clarify my thinking (or given me further pause for thought), with Paul Auerbach, Joe Bailey, John Benetti, Barry Cavell, Mike Chapman, Gail Cunningham, Mike Hawkins, Mark Penson, Roger Pond, Steve Smith, Bridget Towers, Gerry Turvey, Keith Weightman, and Mike Whittingham.

I am also indebted to an anonymous referee, whose sympathetic comments and suggestions have helped me position the argument in a broader and more supportive context of debate.

Special mention is due to Myradh Kinloch, who read portions of the manuscript and without whose clarity, the text would be a lot more cumbersome than it already is. I owe Myradh infinitely more than this, however.

Of course, the errors are mine.

Finally, I must thank the creationists whose writings I have so demonstrably appropriated. My thanks to David Rosevear, the Chairman of CSM, for permission to use their pamphlets in this study and to reproduce portions of them here. Although I may not have taken from these writings the meanings that were intended, I hope the creationists will find some consolation in the fact that they, too, have made me think and question and wonder in ways that I would not have done without them. Surely that is no bad thing?

—*Simon Locke*

ॐ ✧ ☙

Let us ... clarify what this intellectualist rationalization, created by science and by scientifically oriented technology, means practically.

Does it mean that we, today ... have a greater knowledge of the conditions of life under which we exist than has an American Indian or a Hottentot? Hardly. Unless he is a physicist, one who rides on the streetcar has no idea how the car happened to get into motion. And he does not need to know. He is satisfied that he may count on the behavior of the streetcar and he orients his conduct according to expectation; but he knows nothing about what it takes to produce such a car so that it can move .... The increasing intellectualization and rationalization do *not*, therefore, indicate an increased and general knowledge of the conditions under which one lives.

It means something else, namely, the knowledge or belief that if one but wished one *could* learn it any time. Hence, it means that principally there are no mysterious incalculable forces that come into play, but rather that one can, in principle, master all things by calculation. This means that the world is disenchanted

—Max Weber, 1948

There are some things man is not meant to know.

(*The Invisible Ray*, 1936, cited in Woodlief, 1981.)

1

Let There Be Rationalization

And God said, Let there be light: and there was light.

Genesis 1:3

In the final scenes of the 1974 science fiction film *Dark Star,* a malfunctioning, artificially intelligent bomb explodes, destroying the spaceship carrying it. Prior to this, the ship's crew had attempted to persuade the bomb not to explode, by teaching it phenomenology. This involved getting the bomb to question the validity of the command to explode, by asking it to examine its knowledge of the source of the command. Reflecting on this, the bomb shut down, caught in an apparently irresolvable paradox—that the status of the command to explode depends upon sources that appear to be external, but which might not in fact be so, as the only evidence for their externality comes from the detection of signals that are actually internal events. Finally, the bomb arrives at the ultimate point of solipsism; concluding it is the only entity in existence, it assigns itself God-like status and explodes in what it takes to be a Big Bang-like act of creation. As it does so, it says: "Let there be light!"

Dark Star is a witty film. It has been described (Ash, 1977) as a parody of the 1968 highly successful film *2001: A Space Odyssey,* but more accurate is Clute & Nicholls' (1995) broad view of it as "a sophisticated mixture of black comedy and genuine science fiction" (p. 300). The film twists many of the established conventions of science fiction and punctures more optimistic visions of scientific and technological advance, through a focus on human frailty. Added to this is the play made with the relationship between science and religion, summed up in this last act of explosive madness. Crucial to this is the bomb's adoption of the language of God as given in the Bible.

3

In this, the film does two things that comprise the main concerns of this book. First, it provides an example of a more general feature of modern culture: the exploration of the boundary between the scientific and the nonscientific. Science fiction is an especially fertile zone of such exploration, but it is by no means the only one and, beyond this opening illustration, figures no further in this particular book.

Second, the play of this scene draws upon a textual reference, which, it can be assumed, is readily recognizable to a wide audience. The quotation, "Let there be light," together with the range of possible connotations its biblical source might engender, must be sufficiently widely recognized as part of the commonsense knowledge of modernity for the pun to be meaningful to a potential audience of movie-goers. As such, it can be described as commonplace (Billig, 1987) in our culture.

This, I want to argue, is important because it indicates that religion (Christianity, especially) lives on in modern culture, not necessarily or solely as an institutionalized belief or ritualized practical ethic, but as a *resource*, a repository of signification, of meaningful signs, narratives, themes and, most importantly for the present purpose, of *arguments* and the *dilemmas* they are often centered around.

The film, then, can be understood as playing with a philosophical dilemma over the nature of the knowledge of existence, which it takes to be a problem understandable to competent members of the viewing audience (the public at large) and which it resolves by drawing upon theological resources, again assumed to be understandable (but not necessarily believed or accepted) by competent members. Further, this usage is intentionally humorous, assuming that competent members will recognize as absurdly witty the idea that an intelligent machine might come to see itself as God. The film exploits a range of cultural resources drawn from the repositories of modern science and traditional religion and uses them to construct a scenario in which the boundary between the two is—literally and metaphorically—exploded.

Oddly enough, that is also exactly what creation scientists do.

Creation science presents a major puzzle to contemporary sociology. Creation scientists are (mainly) Protestant fundamentalists who reject the theory of evolution in favor of the account of origins set out in the book of Genesis in the Bible. What makes them so puzzling is their claim that this rests on scientific grounds. They claim to be, not just fervent Christian believers, who accept God's Word as truth, but scientifically informed

Christians, who have marshalled a range of evidence and a body of argument against evolutionism and for creationism[1].

Creationism is a puzzle, because it seems to fly in the face of the usual model of modernity adopted by sociologists. In this model, modernity is a scientific society, which is to say, a society in which science plays a central role, not just productively and organizationally, but as a system of knowledge and a basis for the modern outlook on life. The spread of science is not just through the industrial system of production, but also through the cultural order, having profound effects on the consciousness of the modern as much as upon material well being. In essence, science—and the conjunction of reason and materialism that it embodies—progressively displaces traditional beliefs, eclipsing religious, superstitious, and irrational worldviews and knowledge claims (Wilson, 1966). I refer to this model of the unfolding developmental logic of science in modernity as the rationalization hypothesis.

Given this hypothesis, then, what is to be made of creationism? To sharpen the problem, according to common observation, creationism has enjoyed growing interest and public support since the 1960s (Barker, 1979, 1985; Lessl, 1988; Nelkin, 1982, 1992a, 1992b; Numbers, 1987; Toumey, 1994). Much of this is in the United States, and elsewhere, creationism has seen growing support (Numbers, 1987). Even in Britain, where creationism is, at most, a minor public presence, there nonetheless persists what is probably the oldest, continuous existing group of creationists in the world, the Creation Science Movement (CSM), founded as the Evolutionary Protest Movement (EPM) in 1932 (Turner, 1982). So, in just that period of history when science achieved the height of its institutionalized authority, when "Big Science" (Price, 1984) arrived, there has also appeared a movement concerned to restore the credibility of traditional belief.

This puzzle is further complicated for those who perceive a deep contrast between the religious and the scientific, especially in terms of the mode of thought that each entails. This, again, is characteristic of the rationalization

---

[1]For convenience, I use the terms *creation science* and *creationism* interchangeably. Obviously enough, *creationist* is a term that might be applied to a range of versions of Christianity, with varying views about the meaning of the Bible and its relationship to modern science (see, e.g., Toumey, 1991). Equally, it may also be applied to a variety of nonChristian beliefs (e.g., Barker, 1996; Layton, 1996). Meanwhile, the term *creation science* is already a loaded one; using it might be seen as an endorsement of the creationists' claim to be scientists, when this is itself the central concern for many commentators and critics of creationism. Indeed, for many, there is simply no debate—creationism is not science. My free use of the term creation science should not be taken as an endorsement of their views. It is, rather, a sign that, within this study, I treat their discourse with equal credibility as that of their opponents. See the discussion of methodological symmetry later.

hypothesis, which proposes a sharp separation between the mental outlooks or orientations engendered by traditional and scientific worldviews—as seen, for example, in Weber's (1948) description of intellectualist rationalization given previously.

Rationalization has been central to the kinds of explanations of creationism advanced to date within sociology. This book is an attempt to present an alternative view based on a different perspective, that of discourse analysis (conventionally abbreviated as DA). It is also an attempt to explain why such an alternative is needed, as well as to defend the strengths of DA in application to the general field of study of the public understanding of science.

The argument, basically, is as follows. Recent research in the public understanding of science has served to highlight a significant methodological concern, which also carries important theoretical implications for sociological conceptions of the role and position of science in modernity. In fact, these theoretical issues already follow from the development within sociology of the sociology of scientific knowledge (SSK), but they have largely gone unacknowledged. The growing interest in the public understanding of science, however, now lends them a sharper focus (Gross, 1994; Irwin & Wynne, 1996a; Woolgar, 1996; Yearley, 1994).

The methodological difference centers around two senses derived from the word *understanding,* to mean either acceptance or interpretation. One position holds that we should view the public understanding of science as a matter of the level of public acceptance of the knowledge claims made by members of the orthodox scientific community; the other is that we should view it as a matter of the public's interpretations of science and the knowledge claims made by scientists. These two positions carry theoretical implications for sociological conceptions of the role and position of science in modernity. Traditional sociological conceptions of an unfolding logic of science share with the acceptance view certain core assumptions—or representations—of science and the public. SSK, on the other hand, has more in common with the interpretation view—and, within this, the specific focus of interest of DA is on the ways in which science is presented and used within particular social contexts. This leads to a concern with *rhetoric*.

The focus on the public acceptance of science can be said to draw on a similar rhetoric as is found in the rationalization hypothesis, one that represents science monolithically, as a reified, singular thing, in counterpoint to the public also represented monolithically, as a single mass (Irwin & Wynne 1996a; Whitley, 1985). In contrast, the focus on public interpretations of science treats the uniformity of science as a rhetorical accomplishment of scientists and sociologists and anticipates local variations in the characterizations of science between and among public groupings, as an outcome of variable interpretations arising from the discursive and rhetorical resources differentially available to people in different social contexts.

Complicating this simple contrast, however, is the observation that the rhetoric of rationalization is also used by members of the public in their characterizations of science. At least, this is true of creationists. Further, it is also notable that there are systematic variations to be found in the rhetoric of rationalization itself, in terms of its characterizations of science and of the public understanding of science—and, in the case of creationism, of this as well. Additionally, such variations—particularly in the characterization of science—appear in the discourse of creationists also. This suggests that there is *a shared discourse of science* uniting these otherwise contending groups. A key feature of this discourse is the rhetorical reification of science, its representation as a monolithic and stable entity. In fact, however, it is not so stable, as it can be seen to shift and vary in ways that are patterned rhetorically to achieve particular effects in the characterization of an opponent and, thereby, to undermine that opponent's legitimacy.

This kind of observation has been made before in respect of the rhetoric of scientists themselves. Gieryn (1983), for example, in his study of boundary work, showed that members of the scientific community mobilize different, and not altogether consistent, characterizations of science in different contexts to construct boundary demarcations between, for example, science and religion, and science and engineering. Similarly, there have been several examinations of the boundary work undertaken by members of the orthodox scientific community in their attempts to exclude creationism and to undermine the legitimacy of creationists' claims to be scientists (Gieryn, Bevins, & Zehr, 1985; Lessl, 1988; Prelli, 1989a; Taylor, 1992, 1996). With the exception of Taylor, however, much less attention has been paid to the creationists' own boundary work, that is, their turning of the tables on the evolutionist orthodoxy by using criteria of demarcation, in conjunction with a range of other rhetorical and discursive means, to challenge the legitimacy of evolution itself.

Moreover, little of this rhetoric-informed analysis seems to have penetrated into wider sociological accounts, many of which continue to work with an unreflexive monolithic construction of science. It is as an effect of this that creationism appears as a puzzle requiring explanation by the resources of sociological analysis—specifically, those provided by the discourse of rationalization. Yet, creationism only emerges as a puzzle in this form because of the adoption of this discourse. Meanwhile, the creationists themselves adopt similar monolithic representations of science and its impact, using them to undermine evolution and account for its existence in modern times. For them, it is evolution itself that now appears as the puzzle—an atheistic theory in an essentially God-given world. Thus, the discourses of the two groups parallel each other in their shared representations of science and their usage of them to undermine each other's legitimacy, with the seemingly paradoxical consequence that the opponent is in each case constructed as a phenomenon that can only properly be understood and explained by its opposite's own discursive order!

My use of the term *discourse* will be further explained later. For the moment, what needs to be noted is that I treat discourses as *resources* both of representation, and of rhetorical techniques and devices for the purposes of argument designed to persuade others of the adequacy of the representations advanced. These resources are available to people to varying, undetermined extents. Thus, the discourse of science provides a repository of characterizations (of science, of the public, and of other things); modes of expression (such as interpretative repertoires); and means of argumentation (ways of legitimizing and delegitimizing knowledge claims), all of which may be utilized *selectively* by members.

In addition to previous arguments, one further level of theoretical argument is offered. Rationalization can itself be understood as emerging from the attempt to resolve an *internal dilemma* in thinking about the role and position of science in modernity (Locke, 1997). The dilemma is given by the question: How can a universally valid form of knowledge (science) arise from a particular society (modernity)? Given this, I treat rationalization as one discursive instantiation within the broader field of the discourse of science within modernity—that is, as one set of representations and argumentative resources. As will be discussed further later, rationalization offers one resolution of this basic dilemma, but the dilemma itself remains. Therefore, the broader set of discursive resources concerning science also remains potentially available to members, to be drawn upon, where possible or desired, as a means of constructing alternative views of science and its significance.

Overall, then, I return to my basic point, that there is an ongoing argument about science and its boundaries within modernity. Consider again *Dark Star*. The film can be thought of as an example of what Billig (1987) called "witcraft," that is, the ability to display skillful application of the rhetorical resources commonly available within one's culture. In its play with the boundary between science and religion, it draws selectively on characterizations of these and their relationship, skillfully employing them to arrive at the punch line, in which a machine adopts the language of God in a perverse parody of the Creator's act. Further, the film relies on its audience's own *witcraft*—their knowledge about the selectivity of these characterizations—for them to get the joke. Unless they are capable of recognizing these characterizations, *and of knowing them to be deliberately partial and specially chosen*, they are unlikely to appreciate the humor (though whether they find it especially funny or not is another matter). Thus, there is a shared discourse of science within modern culture. However, it is a discourse that is not internally coherent; it is fragmented and filled with tension—hence, *dilemmatic* (Billig et al., 1988). This book is a first attempt to chart the workings of this dilemma, through the application of the DA perspective to the case of creation science.

The rest of this chapter further discusses the methodological and theoretical issues outlined previously. First, there is a brief analysis of the acceptance approach to the public understanding of science, designed to bring out the monolithic representations of science and the public it makes. This is contrasted with the main conclusions to be drawn from research adopting the interpretive approach. This is used to elaborate further on the theoretical difficulties confronting the rationalization hypothesis, considered initially in regard to the attempt to account for the apparent growth of so-called antiscience movements in modernity, of which creationism is often taken as an example. I argue that these terms themselves are engendered by the adoption of the discourse of rationalization, with its monolithic representations of science and the public, specifically in terms of Weber's notion of disenchantment. This position is contrasted with one derived from SSK. Here, I draw in particular on Barnes (1974), whose early statement of SSK incorporated an outline of the notion of *science as a cultural resource*. I develop this preliminary notion further through the perspective of DA, to provide the background for the detailed analysis of creationism expounded in ensuing chapters.

PUBLIC UNDERSTANDING OF SCIENCE

In recent years, a significant gap has been identified in our knowledge of the public meaning of science. This occurred at a time when this issue has come to be seen as being of pressing public importance. There is now widespread concern expressed throughout the academic community, although especially among natural scientists, over the apparent growth of so-called antiscience (e.g., Holton, 1992). At the same time, divisions have emerged within the academic community, in the form of the so-called science wars (Ross, 1996), which appear to reflect more general processes of apparent fragmentation of culture, with what some see, and fear, as an associated loss of authority of science (e.g., Dunbar, 1995).

This raises the question of why the public understanding of science—broadly defined—has been neglected for so long, especially by sociologists. It is only in recent years, following the publication of the report by the Royal Society of London in 1985, that any systematic research into the topic has begun, at least in the United Kingdom[2]. There is, I believe, a straightforward reason for this neglect: Quite simply, it was assumed that there was nothing much in need of researching as it was thought we already knew what the public understanding of science looked like. And for our answers, we turned to Weber.

Weber's (1948) view of the process of "intellectualist rationalization" has become something of a commonplace in contemporary sociology. Primary discussion of this issue has centered around the question of the secularization of the modern world, especially regarding the displacement of one institution of knowledge production (religion) by another (science). However, although agreement over secularization remains far from settled (Hamilton, 1995), the view of 20th century culture as rationalized and disenchanted is nonetheless widely taken for granted. One consequence of this—perhaps also stemming from Weber's (1948) discounting of the relevance of public knowledge of science—is that what is now called the public understanding of science, however defined, has been largely bypassed as an issue of concern. Basically, a hypothesis was accepted largely as a fact, and any process of scientization of modernity that might have occurred was treated more as a given than a topic requiring detailed study. In addition, it can further be suggested that, not only does disenchantment hold no more than hypothetical status as a knowledge claim, but it is itself a discursive representation that can be treated as a *topic* of study, rather than, as Weber takes it, as a *resource* of explanation. I pursue this point further in the next section.

For now, the focus is on the emerging body of research into the public

---

[2]There is a somewhat longer tradition of survey research of scientific literacy in the United States, summarized in Miller, 1983; 1992. See also Wynne, 1995.

understanding of science.[3] As outlined previously, there is a contrast in methodologies dividing much of the research undertaken to date. Elsewhere (Locke, 1994b), I summarized this as a simple contrast in emphasis, between public understanding of *science*, and public *understanding* of science, and I continue here to use italics to signal which sense I am intending. The first emphasis refers to research that is interested in finding out how much members of the public know about accepted scientific knowledge. This research is most clearly represented in the survey studies carried out in the United Kingdom by Durant and his colleagues (Durant, Evans, & Thomas, 1989, 1992; Evans & Durant, 1989; see also footnote 2). I do not intend to spend any time discussing the findings of this research, as I am more interested in the background assumptions on which it draws. Basically, the general conclusion is that public knowledge of science is not all it might be and could be said to confirm Weber's claim that there are more streetcar riders than physicists.

This research has been referred to as working with a "deficit model" (Durant et al., 1992; see also Gross, 1994; Ziman, 1991) of public understanding, meaning that it is constructed with the aim of finding out what the public *does not* know, rather more than what it does. In contrast, research that is interested in public *understanding* of science has a much less restricted scope, concerned to establish what members of the public *make* of science. This goes beyond the simple testing of knowledge (although knowledge claims may be part of the research interest), to consider the meaning that science has for people and the use made of it within specific social contexts (Irwin & Wynne 1996a; Wynne 1991).

An important focus of interest here is on representations of science in the public sphere—the view I develop is that rationalization is one such public rhetoric of science within modernity. Initially, I show that research concerned with the public understanding of *science* draws on a very similar rhetoric, adopting similar kinds of characterizations of science and its social impact. This has implications, not only for how the public and its relation-

---

[3]A number of other developments in related fields also have some bearing on this topic. This is not the place for a detailed outline, but note: the study of science fiction and science fiction fans (Jenkins, 1992; Tulloch & Jenkins, 1995); the consumption of technology (Grint & Woolgar, 1997; Jackson & Moores, 1995; Mackay, 1995; Silverstone & Hirsch, 1992); the cultural significance of the cyborg (Gray, 1995); and challenges to established medicine (Sharma, 1992; Williams & Popay, 1994). Recent interest in "Green" issues (e.g., Yearley, 1991) and, more broadly, the so-called risk society (Beck, 1992) has also fed this focus. Broader still, the tide of interest in the so-called postmodern can be interpreted, in part, as a response to the exhaustion of the rationalization hypothesis. I take this view in my discussion of Lyotard in chapter 6. Other pertinent studies are: Ashworth (1980); Cameron & Edge (1979); Dolby (1996); Hess (1993); Restivo (1978, 1982); Rosenburg (1961, 1972); Turkle (1984); Wallis (1979).

ship to science is represented, but also specific significance for the representation of creation science. It can be suggested, in fact, that this rhetoric is doing boundary work (Gieryn, 1983) for the scientific community. It is debatable, therefore, how much it contributes to the clarification of the public *understanding* of science; arguably, it tells us more about how some members of the scientific community perceive themselves and wish to be perceived publicly. Although this may be a legitimate concern for them, it is right that sociological analysis should strike a different emphasis, as is readily seen from the second strand of research emphasizing the heterogeneity, complexity, and multiplicity of public *understandings* of science. This strand points toward a different theoretical background, in SSK, which has broken away from the rationalization perspective.

The rationalization hypothesis speaks of both science and its cultural impact in reified and monolithic terms—of one singular thing (science) impacting on another singular thing (modern culture) in a uniform way. A very similar outlook can be seen at work in the Royal Society's (1985) report. Significant here is the stress placed on the need for improvement in the public understanding of science. As Wynne (1992a) argued, this stress shows that the report's authors already had in mind a particular sense of what the public understanding of science should be—specifically, that the public should *accept what orthodox science says.* To do otherwise is to be in error. Thus, the report mobilizes a model of orthodox science as *internally* unified, on the basis of which the claim that there should be a similar uniformity *externally* can be justified.

Although already the subject of some criticism (Irwin & Wynne, 1996a; Michael, 1996a; Wynne, 1992a), it is worth looking here in a little more detail at the rhetoric of improvement in this report, in order to bring these points out fully.

One notable feature is the manner in which the concern with the limited knowledge about the public understanding of science is constructed. The report's authors state that: "There are many surveys of *attitudes to* science and technology both in the UK and overseas .... But there has been much less effort outside the formal education system devoted to assessing the *understanding* of science and technology." (Royal Society, 1985, p. 12; emphases original).

The use of the word *assessing* here is telling, with its familiar educational connotation of *testing.* This might seem quite an appropriate term, then, to be used for testing the public's knowledge of science. However, the word is not without an ambiguity, familiar to every student and teacher. Within education, especially, there is a common dilemma experienced about assessments, whether they are concerned just with finding out about, or, more menacingly, with finding out. To assess in the first sense means to observe, discover, and become informed; but, to assess in the second sense means more to find fault and to disclose areas of weakness, inadequacy, or error (to try the value of, in the dictionary meaning). Thus, however much the sympathetic teacher may seek to reassure students that the concern is only to find out what they know, the shadow of a sterner judgment hangs over them nonetheless. There is also reason to think that it is on the services of this sterner judge that the Royal Society are principally calling.

The previous quotation serves two points simultaneously: It is declaring that we do not know much about the public *understanding* of science and that, given its importance, we ought to do more to find out about it; and it also implicitly mobilizes a measuring rod, saying that we should find it out. As such, it is a claim made in the expectation of finding inadequacy. Moreover, the statement was made within a context that has been predefined as one of a need for improvement (Royal Society, 1985). This serves to authorize (Smith, 1978) the particular reading of the text as presupposing error.

The statement, then, is a way of formulating our state of ignorance about the public *understanding* of science, that directs our attention toward a concern with the public understanding of *science*. It uses the ambiguity in the notion of assessing to encourage us to adopt the perspective of the stern teacher who wishes to ensure that lessons are properly learned and designs tests accordingly. It is this stern attitude, we shall see, that has become embodied in the social survey questionnaires favored by the deficit model.

Additional features of the Royal Society report of interest include the argumentative strategies used to justify the concern with improvement. First, reference was made to general levels of literacy and numeracy, where it was argued that the inability of a proportion of the population to be able to read or count excluded them from developing much grasp of science. However self-evident this might seem, it is nonetheless important to recognize that it advances a particular sense of the nature of scientific knowledge—as codified—and as something to be arrived at in a specific kind of way—from books. This connects readily with the view that a proper understanding of science can only be arrived at through a form of institutionalized training. Thus, in this respect, a defense of current institutional structures of scientific learning is implicit. Understanding science, it would seem, is not just about the quality of knowledge, but also about how—and from whom—it is obtained. This has the effect of excluding other routes to knowledge, ones which, arguably, are far more likely to be used by lay members to obtain knowledge of science. For example, it excludes arriving at scientific understanding through practical experience[4], or, by implication, through other media of knowledge transmission. It also excludes alternative knowledges and understandings (i.e., interpretations) to those that bear the official sanction of the current scientific community as imparted through the mechanisms of professional and occupational socialization.

Given this, it is unsurprising that only these other processes were singled out as cause for concern by the authors of the report. Thus, for example, they observed the "considerable public *interest in* science revealed by the findings of ... opinion surveys ... and by the proliferation of popular technical magazines and hobbies" (Royal Society, 1985, p. 15; emphasis original). However, rather than assess the quality of public knowledge on this basis, they preferred instead to gather opinion from "individuals or organizations *professionally* involved with science" (Royal Society, 1985, p. 15; My emphasis).Thus, professional opinion was favored over the public's self-chosen activity. Moreover, this opinion was found to be negative: "Not surprisingly," it was said, "this evidence unanimously argued ... that public *understanding* of science was inadequate" (Royal Society, 1985, p. 15; emphasis original). The interesting question here, however, is why this particular finding should be considered not surprising. It is tempting to see in this a concern less with what the public may actually know (and think) about science and more with controlling what they know and from where they get their knowledge.

This is especially evident in the concern expressed about pseudoscience. There is an interesting contrast to be observed here. One of the findings from professional scientists is a worry that the public have a rather simplistic view of science as a "purveyor of certitudes," providing "unequivocal answers" (Royal Society, 1985, p. 15). However, the authors also suggested that, "Greater familiarity with the nature and the findings of science will ... help the individual to resist pseudo-scientific (sic) information ...." (Royal Society, 1985, p. 10). Much of this pseudo-science, the authors seemed to believe, is conveyed through the mass media. The media were said to be subliminal sources of information (Royal Society, 1985); although no

---

[4]It is interesting to note here the kind of view of science advanced by Dunbar (1995), or indeed, Popper, in his later works (e.g., Popper, 1972), in which it is presented as being, in part, knowledge founded on practical experience of dealing with the world. Admittedly, Dunbar also wanted to emphasize the importance of mathematics to modern science, but this serves nonetheless to highlight the awkwardness of attempting to justify science as both a universal form of knowledge and the special province of a hierarchy of experts. This problem is explored further later.

evidence in support of this claim is offered. Similarly, it was claimed that: "some popular books ... are grossly inaccurate and even positively misleading .... [S]ome science fiction can grossly distort scientific possibilities and create much concern in a public with limited scientific literacy and so limited ability ... to sort the plausible from the implausible or rank impossible" (Royal Society, 1985, p. 23).

Here, then, there is the strong sense that science is able to purvey certitudes, in the sense that it can at the very least tell us what is and is not possible.

Also of note here is the use of an extreme case formulation (Pomerantz, 1986), which imparts a particular kind of stress to this judgement. The phrase "rank impossible" appears as the final item in a three-part list (Jefferson, 1991) beginning with "the plausible" and moving to "the implausible" before ending with this "rank impossible." Three-part listing seems to be a widely available discursive technique, generally used "to indicate a general quality common to the items in the list" (Wooffitt, 1992). As such, it conveys a sense of completeness: Listing three items seems to work to suggest that all possible items of relevance are covered. In the present instance, however, the list seems to be employed not only to convey a sense of completeness, but also to enhance a sense of difference and exclusion—that is, to construct a demarcation.

The list covers all possible ways of describing (apparent) phenomena, or the claims made about them: They may be plausible, implausible, or impossible. It also constructs a progressive boundary of acceptability, a boundary given an added rhetorical charge by the use of the word *rank*. In describing the impossible as rank, it is located as far away from the arena of plausibility as possible; moreover, it is also invested with a sense of strong negative judgement. Thus, the movement from the plausible to the rank impossible is not just a movement from what is to what cannot be; it is also a movement from what is to what ought not to be. This is a rank of valuing as much as of being. When things are rank, they are bad; more precisely, they are gross, atrocious, and even accursed (synonyms from *Roget's Thesaurus*). Thus, popular, fictional, and pseudoscientific representations of science are not merely wrong, they are a corruption—and it is worth noting that creationism in particular has been described as an abuse by at least one defender of scientific orthodoxy (Kitcher, 1982), and Lessl (1988) has likened the response to creationism as akin to an orthodox Church responding to heresy.

Overall, then, what can be said about the Royal Society's report is that it is designed to construct the factuality (Potter, 1996; Smith, 1978) of public ignorance and misunderstanding of science, thereby justifying the demand for improvement of the situation. Implicit in what is said is a normative ideal of what science itself is, and what public understanding should be like. Science is a certain type of knowledge (codified), obtained in a certain type of way (from professional training), and readily distinguishable in a once and for all fashion from things that are dressed up to look like science but are not (adverts, pseudo-science and fiction). Further, the opinions of professional scientists are a better judge of public understandings than the activities of the public themselves, and, despite the evident widespread interest in science, the public are uninformed and vulnerable, subject to propaganda from the media, including subliminal messages about pseudoscience, which they are ill-equipped to deal with competently, because they have limited scientific literacy (Royal Society, 1985). Overall, it appears they are in a dreadful mess and a cause of major concern, as it is not just science that is at stake, but the future of industrial society and, indeed, democracy itself (Durant, 1990; Royal Society, 1985). Little wonder Fuller (1997) described the appearance of this sudden concern with public understanding as a "moral panic."

Thus, there is only *one* science and there is only *one* correct understanding. Significant here especially, is the implicit exclusion of any alternative, any argument or criticism of orthodox science will appear as misunderstanding or misrepresentation. The construction is a closed circle: To understand science means choosing to believe what orthodox scientists say is the case; if, for any reason, one chooses to reject orthodoxy, one has misunderstood —it is relegated to ignorance, propaganda, illiteracy, or some other error. Interestingly, as presented in chapter 4, it is precisely to factors of this nature that *creationists* also ascribe the broad acceptance of evolution within contemporary society. These factors are part of a range of devices that constitute a common rhetoric of "accounting for error" (Gilbert & Mulkay, 1984; Mulkay & Gilbert, 1982).

As the label might suggest, the rhetoric of error has been taken up also in the deficit model of research in public understanding. As this model has been the subject of some criticism elsewhere (Bauer & Schoon, 1993; Locke, 1994a; Wynne, 1991), I do not wish to devote much space to it here. There are, however, two points relevant to the current study that need to be made.

First, this type of approach effectively denies the possibility that members of the public may have an understanding of science as involving disagreement and argument. As Bauer and Schoon (1993) argued, these surveys effectively reify a Popperian model of science, involving the empirical testing of hypotheses deductively derived from theoretical statements. However, even Popper (1952) recognized the importance of argument and critique in the workings of the scientific community. These surveys effectively denied the public's capacity to present an understanding of science of this form.

Further, Durant et al. (1992) asserted that, for the purposes of testing public understanding, science may be treated "as a reasonably stable body of knowledge," about which there is agreement by "all competent experts" (p. 163). These can be used as a yardstick of public knowledge. This is interesting to note, because among the criticisms of creationists is that they mistakenly treat science as a body of knowledge, or "a collection of facts" (Nelkin, 1982; see chapter 3 for further discussion). This is taken as evidence of their misunderstanding of science. Again, this bolsters the view that these claims serve rhetorical ends.

The second point is one I have made elsewhere (Locke 1994b), but needs brief restating here because of its particular relevance to creationism. One aspect of Evans & Durant's (1989) survey specifically dealt with the public acceptance of science. To test this, they included some questions that were designed to determine to what extent people accepted the Big Bang theory of the origins of the universe and the theory of evolution. The thinking behind this relates directly to their recognition of the growth of creationism in the United States and an interest in its presence in the United Kingdom. Specifically, they refer to "religious fundamentalists [who] have opposed scientific theories of human origins for the greater part of the twentieth century" (Evans & Durant, 1989, p. 114).

This discussion is set against a general reference to public skepticism about "the findings of science for what are *essentially ideological* reasons" (Evans & Durant, 1989, p. 114, My emphasis). In other words, they establish a boundary between science and nonscience (ideology, and specifically religious fundamentalism) as the basis on which to measure acceptance of science. In so doing, any claim to legitimacy that might be made for the creationist viewpoint is already undermined; it is methodologically predetermined as outside the boundary of the scientific. This does more than merely see science as a body of agreed knowledge; it also draws on a rhetoric that asserts a hard and fast distinction between science and nonscience ideology, which implies that it is possible to position beliefs, theories, and frameworks of thought clearly and definitively in either one category or the other. Once again, then, this makes it effectively impossible to measure the extent to which public understanding of science involves recognition of the role of disagreement and argument within science on this type of issue. There is an assumption of stability and uniformity within the scientific community, and of the clear demarcation of science from everything else.

However, in the case of creationism (as well as many other things) there is good reason to think this inadequate. For example, not only do many creationists claim to be fully qualified scientists (Lessl, 1988; Nelkin, 1982), but, creation scientists, in general, do not present themselves as rejecting science, only evolution. There is no way this can become clear from Evans & Durant's method, because they effectively defined science as coterminous with evolution; to reject the latter, then, is automatically to reject the former. This, then turns the problem of public understanding of science into one of a competition over labels and the right to ascribe them to particular discourses represented as monolithic objects—this one is science, that one not. This does not allow for the possibility that members of the public may be predisposed to adopt alternative characterizations of science, or, as is perhaps more likely, to recognize the existence of different characterizations and to share in the uncertainty over their applicability that is often found among academics themselves. Thus, it is unclear that we actually learn very much about *public* understanding of science from such surveys. Arguably, they may tell us something about the public understanding of *science*, but perhaps what they mainly tell us about is the researchers' own boundary work.

That there are alternative rhetorics of science in modernity can readily be seen from popular culture. I suggested that the deficit model draws on one rhetoric sharing certain key representations with rationalization, regarding the uniformity of science and its impact on society (in the case of the deficit model, it is clearly held that this impact should be uniform, even if the worry is that currently it is not). Thus, we now reconsider the contrast between Weber's claim about disenchantment and the quotation from *The Invisible Ray*. Weber's claim about the modern mind-set—one that accepts that "one can, in principle, master all things by calculation"—stands somewhat in contrast to that commonplace of 20th-century science fiction that "there are some things man is not meant to know."

This brings out the deep ambivalence (Handlin, 1965) about science within popular culture. As much as people may be impressed by the cold logic of the scientific machine—as also, again somewhat in contrast perhaps, by its "wow" factor—so may they also be concerned about the implications of its advance (cf. LaFollette, 1990). This suggests, for the modern, it is not so much one side or the other of this dilemma that is constitutive, but the dilemma itself. It is the argument, the tension, that makes the story of science, not the progressive unfolding of only one side. Assuming this to be the case, then, it would alert us to the likely presence of variation in the understanding(s) of science within public contexts. Support comes from the strand of research concerned with public *understanding* of science.

Perhaps the most developed branch of this research is concerned with perceptions of risk. I do not propose to attempt any kind of summary of this work, just to note the broad general conclusions that may be drawn (for fuller discussions, see Irwin, 1995; Irwin & Wynne, 1996b). In my view, there are three basic points.

First, in the assessment of risk involving science and technology, people choose from a range of arguments in arriving at their assessments. These arguments may be as much moral, as factual or technical in their grounding (e.g., Hornig, 1993).

Second, in the negotiation of local developments involving science and technology, Nelkin (1992b) argued that people utilize scientific and technical arguments in a tactical fashion, such that technical expertise is mobilized as "a crucial political resource ...." (p. xxi). In particular, as Mulkay (1979) also argued, people often display skepticism about scientific and technical knowledge-claims and are able to mobilize a critique that sees them as politically interested or value-laden. This kind of critique is common in creationist texts. (It should also be noted, however, that, despite the observations made here, Nelkin (1982) nonetheless drew her model of science in the monolithic terms of rationalization, as is discussed further in chapter 2).

Third, all this may be summarized in the view expressed clearly by Wynne (1992b) that public understandings are likely to be highly complex, working on a number of different levels concurrently, some of which might be represented as acceptance of science and some as rejection, but with an overall "deep ambivalence and apparent inconsistency." In his view, this reflects the "multiplex, not necessarily coherent, dimensions of social identities expressed in interleaved social networks and experiences" (p. 299). As an illustration, Wynne (1992a; cf. Michael, 1992) gives the case of Sellafield workers, who operate with a deliberate socially constructed ignorance of the science behind their highly technological work environment. The workers accounted for this as due to a concern that more than a practical working knowledge of the technology might inhibit their effectiveness in doing the job, through increasing their awareness of the possible dangers. A further illustration is the Cumbrian shepherds, whose skepticism about official scientific views in the wake of

Chernobyl stemmed from the shepherds' belief that the scientists' claims were made in ignorance of local farming conditions (Wynne, 1992b; see also Paine, 1992). The shepherds' knowledge, significantly, was derived from years of experience, not from library books.

These conclusions point to multiplicity and variation in public perceptions and interpretations of science, with related theoretical and methodological implications. Methodologically, it is clear that the deficit model is unable to capture this kind of variation. In adopting a uniform view of science and in focusing its concern on the public's knowledge of this view, responses must suit the predetermined categories to qualify as acceptable, or else be judged wrong. If our interest is in capturing variation, therefore, we must clearly use other interpretive methods. Moreover, if it is the case that the deficit model shares common ground with the rationalization hypothesis, then the theoretical implications of this need to be considered and are considered next.

LET THERE BE RATIONALIZATION

The reified view of science as a monolithic entity standing somehow outside of, but also centrally located within, society is usually associated with the Enlightenment. This is sometimes contrasted with the alternative view of science developed from the sociology of knowledge (e.g., Irwin, 1995). However, it is possible to attribute both of these perspectives to the Enlightenment inheritance. Following Billig et al. (1988), this inheritance can be seen as dilemmatical. In the case of science specifically, I argue, it has bequeathed, not a solution to the role and position of science in modernity, but a problem: How is it that a particular society (modernity) has produced a universal form of knowledge (science)?

It is of course the case that, in this formulation of the problem, a monolithic view of science is advanced, presenting it as a universally valid, and therefore unchanging constant. Nonetheless, that there is a dilemma here is apparent from the discourse of sociology itself, which, in its development out of the Enlightenment background, has increasingly come to articulate a dualistic understanding of science, as both a reified, asocial, universal monolith, and as a socially grounded, contingent, and particular form of knowledge. Further, this contrast is not just sociological, but is also to be found articulated in various ways throughout the commonsense lifeworld of modernity, in the form of ongoing arguments about the nature and meaning of science and its products.

A notable feature of the early sociological theorists was their concern to distinguish the truly scientific from the merely ideological (Larrain, 1979), in the hope of bringing to fulfillment the Enlightenment promise of a fully scientific society. In so doing, however, they began to articulate the dilemma of science in its intellectualized expression. Thus, science was perceived as having universal validity, whereas ideology was merely a reflection of the partial, or subjective, interests of particular groups or individuals. However, in linking cognitive representations directly to social conditions (Durkheim, 1964, 1976; Marx & Engels, 1964), these theorists not only provided a means of grounding ideological consciousness in social conditions, but opened up the question of what kinds of social circumstances were needed to generate a universal knowledge.

For Marx and Durkheim, this was less a problem of direct concern, because they accepted unreflexively the reified version of science as a universal form of knowledge, seeking only to identify the social conditions for its furtherance in all dimensions of life. That they, nonetheless, articulated the problem of the sociology of knowledge despite this commitment is revealing. It shows that it is the dilemma that is constitutive of modern discourse, not the universal, monolithic version of science. From this, it is then no surprise to find that this same dilemma has become, over the course of the 20th century, sharpened, both inside and outside the university.

But, precisely because it is a dilemma, it has led in two directions at once. Specifically, it has led, on the one hand, to Weber's more or less self-conscious effort at resolution; and, on the other, to a fuller articulation of the counterpoint to the universalistic version of science, developed in the sociology of knowledge and its offshoot, SSK.

Weber's rationalization hypothesis marked a significant break with Marx and Durkheim in one key respect: Whereas they saw the scientific society as a promise for the future, Weber saw it as an already accomplished reality in its most important respects. Thus, although others saw the task of sociology as one of defining the social conditions whereby the scientific society might be achieved, Weber saw it as attempting to understand the historical circumstances that have conspired to bring about this unique civilization and to consider its future implications.

At another level, it can be suggested that, although the dilemma of science in modernity only nestles in the sociologies of Marx and Durkheim, in Weber's it is fully fledged. There is no room here to expand fully on this. Weber's notion of rationalization covers a considerable range of historical and comparative analyses, together providing a view of the modern West as arising from an unprecedented confluence of congruent organizational forms (bureaucracy; the capitalist market; Roman law and its notion of citizenship) and cultural traditions (a universalist religious ethic—Protestantism; Grecian conceptual reflection and an empirical experimentalist art combining to produce science), combining in a sort of self-fuelling momentum toward the maximal realization of the inherent rational tendencies in each of these spheres (Weber, 1976; also, Brubaker, 1984; Habermas, 1984; Parsons, 1968). The key feature for present concerns is Weber's perception of Western, rationalized society as one in which purposive or instrumental forms of action have progressively begun to displace moral, or value-directed forms of action (Weber, 1948). For convenience, I refer to these forms of action, after Habermas, as technical and communicative, respectively.

This is central to Weber's resolution of the dilemma of science in modernity. Basically, his argument is that science can be considered a universal form of knowledge, because it ideal-typifies a form of technical action that has the sole goal of production of knowledge about the empirical world for its own sake. As such, it asymptotically tends toward the substantive realization of formally rational action and therefore embodies principles of rationality that inform all types of social action. Consequently, science can make a claim to be a universal form of knowledge, insofar as it is a product of this kind of technical action. And this type of technical action has itself become realizable within modernity, because of certain accidents of history converging to produce the process of rationalization. There is, then, for Weber, an intimate interconnection between science and modernity understood in terms of the progressive domination of a particular type of action and mode of consciousness, characterized by its technical form. This leads him to the notion of intellectualist rationalization, or disenchantment.

It is not my intention to debate the adequacy of this proposed resolution in relation to Weber himself. Rather, I show how it has developed into more recent theorizations of the role and position of science in modernity. In particular, I highlight the apparent problems that result in the attempt to make sense of certain features of post-Second World War Western culture. Here, I deal with these problems in general; in the next chapter, I focus on the specific case of creationism.

I propose that the notion of disenchantment has entered contemporary sociological discourse as a rhetorical characterization of the role and position of science in modernity. In Weber's theory, as previously argued, it should be seen as a hypothesis, that is, a claim about the empirical world, but not (yet) a description of same. Arguably, however, this claim has been accorded a different status within orthodox sociology, as less of a (tentative) hypothesis and more of a rhetorical commonplace. Disenchantment is treated as a resource to be used for the construction of particular types of accounts, explanations, and arguments; less as one possible way of characterizing the nature of modern consciousness, and more as the way this consciousness is. This generates problems reflecting the partiality of this rhetoric.

I contend that disenchantment is a rhetoric employed by sociologists and other groups and individuals within modern society, and, from this, that it is a commonsensical commonplace. This seems to be somewhat reminiscent of Cavanaugh's (1985) notion of an "empiricist folk epistemology," rooted in a combination of Baconian induction and Reid's commonsense philosophy. He uses this, as have others (Taylor, 1992, 1996; Toumey, 1994), as a means of accounting for creationists' adoption of a model of science grounded in a simple inductionism and commonsense empiricism. However, although there is much to support this view of creationism, it is also the case, as explored further in chapter 2, that this is by no means the only model of science they draw upon. Similarly, in chapter 4, it is shown that creationists do sometimes draw on the language of disenchantment as one way of accounting for the success of evolution. What needs to be stressed, however, is that this is only one of a range of accounting techniques they use. This lends support to the view that disenchantment is a sort of ironic (or reflexive) self-characterization that moderns may utilize as a resource of accounting. Here, as elsewhere, then, orthodox sociology is guilty of conflating its topics and its resources (Bittner, 1974; Boden, 1994).

Further support comes from the confusions that arise within those accounts that employ this rhetoric in attempting to make sense of the public understanding of science. This is especially apparent in attempts to account for cultural movements that are characterized as criticizing or rejecting science. Although I deal with these issues more extensively in later chapters, it is helpful to provide some illustration of these kinds of problems. Although these accounts draw upon a reified characterization of science as a monolithic thing that impacts upon modern consciousness and culture in a uniform and unified way, the attempt to account for antiscience leads to the introduction of arbitrary divisions, whether within the representation of culture or of science itself. These divisions have to be arbitrary, because the discourse of rationalization does not allow for anything other than monolithic representations. The fact of their fragmentation, then, emphasizes their inadequacy.

Two sociological accounts, those of Gellner and Merton, are considered next. These discussions are not comprehensive, but are merely intended to highlight certain features, which I see as symptomatic of the failings of rationalization discourse as an explanatory system (further elaborated in chapter 6). Within this discourse, the effort to account for antiscience produces an unwieldy complexity, in which antiscience becomes not really antiscience at all, but is either a disguised acceptance of proper science, or based upon a misunderstanding of same. Note, however, that this remains an all-or-nothing determination—people accept science, period, or they reject science because they misunderstand it, period. There is no room for variation here, for degrees of acceptance or rejection—for, say, a considered decision to reject the theory of evolution, but not the rest of scientific orthodoxy. This follows from the assumption of uniformity of impact of science on the wider culture. If it is assumed that people are disenchanted, then it follows that, even if they appear to be rejecting science, they cannot really be, because they are "scientized", saturated with the general outlook of the scientific frame of mind, even if they appear to be behaving in a different way. It is impossible for people to reject science in the modern world, because modernity is, by definition, scientific.

A particularly good example of this type of account is provided by Gellner. Two of his earlier books are important: *Thought and change* (1964), and *Legitimation of belief* (1974), in which his understanding of modernity and the role of science within it is set out in most detail (for a more recent statement, see Gellner, 1992b, pp. 136–156).

Thought and change provides a description of the peculiar state of modern consciousness as Gellner sees it. This state is attributable to the disenchantment process of which science is both cause and expression. Science is "the chief single factor" (p. 65) responsible for the great social and economic transformation that produced industrial society and is "both the source and fruit of industrial organisation" (p. 72). Thus, "Modern science is inconceivable outside an industrial society: but modern industrial society is equally inconceivable without modern science. Roughly, science is the mode of cognition of industrial society, and industry is the ecology of science." (p. 179)

This representation of the relation between science and modernity might be taken as characteristic of the Enlightenment ideal. However, Gellner's post-Weberian understanding recognizes that the benefits of science have not come without a price and that the rational society is not won quite so easily. The price of science is radical doubt. The disenchanted consciousness of the modern is one in which questions of morality and value are severed from questions of truth. Indeed, this accounts for the success of science; it is only when freed from the demand to conform to requirements set by a prevailing moral and normative system, that the capacity to arrive at an unrestricted understanding of the natural world is possible. However, such an understanding must also call into question the grounding of the normative system itself. It is not only given natural truths, but social truths as well that are suspended. As a result, the normal condition of modern consciousness is not the state of ultimate truth promised by the Enlightenment, but a state of "controlled doubt and irony" (Gellner, 1964, p. 81).

This is, in many ways, a fascinating and compelling analysis. However, it masks an ambiguity generated by the rhetoric of disenchantment. This can be seen from Gellner's choice of characters taken to represent the modern condition of radical doubt. He referred to the great early modern philosophers—Descartes, Hume, and Kant—but the effect of the reified categories of his discourse suggested that this condition is characteristic of all members of modern society. Gellner spoke not just of these great men, but of all moderns; it is the modern mind, that is doubtfully disenchanted—and, as such, the mind of no one in particular, but everyone in general.

In the language of disenchantment, then, all moderns are philosophical doubters, but only some are taken as representative. This ambiguity remains implicit in the discourse, but worked its way to the surface in Gellner's (1974) analysis of what he calls "ironic cultures," referring to the "new wave of antinomianism, accompanied by emotive, disorderly, near-incoherent doctrines ... [of] the rebellious young of the 1960s" (p. 192). Gellner (1974) rejected the view that these movements herald "an antitechnical, sensualist–mystical, emotive and intellectually undisciplined form of life." Indeed, it is precisely because of their continuing dependence on the underlying productive and administrative technologies that he referred to these movements as "ironic": The irony is that they reject the very technological and scientific progress that makes their way of life possible. Their alternative to the technological–scientific order is, Gellner (1974) claimed, confined to "the residual sphere" of culture.

The ambiguity becomes apparent here in the form of a distinction forged between two strands of modern culture: The "concepts which are part of the serious business of real knowledge"—science; "and the many other styles, whose virtues are different—such as to be jolly, entertaining, homely, or comforting" (Gellner, 1974, p. 193). The "serious business" of science showed its effectiveness to such a vast extent that the very processes it set in motion and continues to govern—economic production, bureaucratic administration, technological development, and so on—have assured a level of affluence "so fabulously high" (Gellner, 1974) that many are able to opt out of some of the marginal benefits. This permits "any degree of fantasy in those aspects of life which are distinct from the serious business of knowledge. The disenchantment of cognitive life, together with the severe separation of thought from play, can actually liberate the latter from constraint." (Gellner, 1974, p. 193)

The growth of this luxuriant culture is no surprise, however, as it follows directly from the fact that science destroys the closed world of tradition, but replaces it with nothing. Nonetheless, some kind of culture—a normative and moral order—is necessary to society. However, it no longer has the same status: "When serious issues are at stake—such as the production of wealth, or the maintenance of health—we want and expect real knowledge. But when choosing our menu or our rituals, we return to culture and religion." (Gellner, 1974, pp. 194–195)

From this, we see the ambiguity in the rhetoric of disenchantment made apparent. The problem for Gellner is that modern consciousness is both of two things at the same time—radically doubting, but uncritical of science. It has to be uncritical of science, precisely because, in his view, it owes its existence to science—in a sense, it *is* science, which is also the fount of material well being in the form of economic and industrial production. Thus, he has to interpret criticism of science as itself a product of science and, therefore, as not really criticism at all.

This applies also to arguments within social science that might appear to be critical of science. Gellner was a notorious critic of relativism and certain philosophical schools associated with it, such as Wittgenstein (Gellner, 1964), phenomenology (Gellner, 1964, 1974), and finally, postmodernism (Gellner, 1992a). In his view, criticism of science, whether of the validity or acceptability of its form or content, cannot be accepted, because critique is effectively identified with science. The effect of this, however, is to dismiss public criticism or rejection of science as meaningless fantasy, as not serious.

In a similar fashion to the authors of the Royal Society report, Gellner constructed public rejection of science in conformity with his own predisposition to its acceptance. Thus, to account for this type of public understanding, he had to introduce what is effectively an arbitrary division into modern consciousness, in the form of the distinction between the two strands of culture. One strand has, somehow, remained insulated (Gellner, 1974) from the other. This has to mean that the radical doubt supposed to define modern consciousness in all spheres—moral and cognitive—has failed to penetrate fully. Yet, how can this be? Why should it fail to penetrate here, if it has penetrated in other places? Moreover, if it is fundamental to modern consciousness to be skeptical and lost in the mists of doubt, then how can it be that some individuals appear to be able to suspend their doubts, to accept the "emotive, disorderly, near-incoherent doctrines" Gellner described? Why do they not find these doctrines as unacceptable as Gellner obviously did? Why are there all these unaccept-able relativists sitting in the universities?

These problems were generated for Gellner by the ambiguities in the discourse of rationalization, which, on the one hand, wants to propose a condition of radical doubt, but on the other, wants to place limits on this condition. Parallelling this, rationalization wants to propose that all mod-erns are disenchanted, but is then left struggling to account for apparent variations in the degree or form that this takes. In order to do so, ad hoc modifications must be introduced into the account, in the form of arbitrary divisions. In Gellner's case, he opted to divide culture, in order to preserve the centrality of science; Merton, on the other hand, sought resolution through dividing science itself.

Merton's writings are valuable to consider both for his way of resolving the problem of rationalization and for his sociological account of science itself, which offers a resolution of the more fundamental dilemma of how a universal form of knowledge has arisen from a particular society. Fa-mously, he resolved this latter problem by developing a model of the norms of the scientific community, which are said to provide a guarantee of universal validity, through making knowledge-claims publicly available to critical scrutiny by suitably qualified peers (Merton, 1968a). This commu-nicative model of science has been widely influential, but also heavily criticized, especially from within SSK. Some aspects of this critique are considered later.

Here, I focus on Merton's (1968b) treatment of social groups among the nonscientific public who express an apparently critical attitude to science, especially those advocating a humanitarian moral stance (cf. Snow, 1964). In some respects, the problem for Merton with respect to such groups is comparable to Gellner's. Just as Gellner saw the modern normative order as given by science, so Merton, drawing, if not uncritically, on a back-ground of Parsonsian systems theory (Parsons, 1951, 1966), presented science as an embodiment of the central value system of modernity (see also, Storer, 1966). This was made explicit in his interlinking of science with a type of social order defined as democratic, effectively conflated with the social and political order of the modern West (and the United States in particular), through contrast with such totalitarian political systems as Nazi Germany and the former Soviet Union (cf. Popper, 1952, 1957). In these latter societies, scientific thought is restricted by state control, but in the West, the critical norms of the scientific community fortunately coincide with and mutually reinforce those of democratic politics (Merton, 1968a, 1968b).

However, in linking science so directly to the central value system of modernity, Merton was faced with a problem similar to Gellner's in accounting for public criticism of science. He was obliged to assume that such public criticism must nonetheless still embody and express the central, scientific values of modernity. Unlike Gellner, however, Merton's solution was to forge a division between the utilitarian (i.e., technical) applications of science, in military and industrial contexts, and its communicative essence. He argued that, although science became associated with these utilitarian interests as a major source of funding in its early development (Merton, 1970), its true nature resides beyond such technical concerns. Ultimately, in its core values, science expresses democratic and humanistic ideals of universal reason, which surpass any narrow or immediate interests. However, in the mind of the lay public, who themselves articulate the core democratic and humanistic ideals of modernity, science remains linked to the short-term interests of a powerful few.

In this analysis, Merton provided a basic strategy that has since been widely adopted as a means of resolving a problem in dealing with contemporary public criticism of science. The problem is that, often, sociological commentators appear to want to agree with such criticism, but do not want to be seen to be rejecting science. (This is not a problem with respect to creationism, by the way, as no sociologist wants to be seen to be agreeing with this!) Merton's solution was to divide science in two, identifying only as real, essential science, those features of scientific practice with which he was comfortable; features with which he was not so comfortable were then open for categorization as nonscience and thereby delegitimized. Further, it is often the mobilization of the category *technical,* or its equivalent, that is employed to accomplish this delegitimation, as can be seen in some more recent variants on this position, such as those of Habermas, Bell, and Lyotard, discussed in more detail in chapter 6.

Although this dissociation of science from the technical might be taken as something of a reversal of Weber's position, what remains in place is the basic discourse of rationalization with its assumption of a monolithic science and a monolithic public, understood in Merton's case through the notion of a central value system. The substitution of a communicative model of science for Weber's technical model is a means of maintaining the integrity of the rationalization framework in the face of ostensibly disconfirming evidence. Rationalization attempts to equate modernity with science, which entails representing the wider social and cultural order as somehow conforming with the logic of science, even if science is, apparently, being criticized or rejected. Merton's solution was to propose that the rejection is based on a *misunderstanding* of science; that it is inadequately informed about and misrepresenting of science and, therefore, directed at the wrong target.

In this, we can see Merton's position allied with that of the Royal Society—although whether they would accept his communicative model of science as well is open to speculation. As far as public understanding goes, however, both apparently propose a view of widespread misunderstanding of science, which is at the base of public criticism and rejection of science. For Merton also, therefore, there seems to be no sense in which public antipathy to science itself (as opposed to some of its uses) can be seen as valid, or as the outcome of considered reflection and active choice. It cannot be, because the critique itself draws on the same central communicative values that inform the essence of science and must therefore affirm it.

Although this by no means exhausts the discussion of the discourse of rationalization, clearly ambiguities and confusions are generated as a result of the attempt to understand apparent criticism of science from the public. Gellner and Merton provided illustration of two types of strategy used to account for public criticism of science, strategies that are also to be found working in different ways into other, more recent, accounts of science in modernity (and postmodernity). Examples of these later developments are considered after the analysis of creationist discourse, as this will enable me to show more effectively that the same basic dilemma emerges in both sets of writings.

Some indication of this, in the case of the sociologists, is already apparent from Merton. The reappearance in his writings of a model of science based on normative conventions, as opposed to the Weberian reduction, is symptomatic of the dilemma of science in modernity. Merton's model of science points away from the idea of a universal formalism of technical action as found in Weber, toward the communicative grounding of science in local contexts of argumentation. However, although Merton might point, he does not travel far in this direction, preferring instead to maintain a grip on the universal through the comparison with consensual political processes. That this model of conventional universalism is not altogether comfortable has become apparent from the critique of Merton from within SSK, considered here next.

SCIENCE AS A CULTURAL RESOURCE

Previously, I argued that the rationalization hypothesis stems from one side of the dilemma of science in modernity and can be understood as a projected resolution of this dilemma. The dilemma is expressed in the question of how a universally valid knowledge may arise from a particular type of society. Rationalization resolves this by effectively defining science and modernity as identical—the one depends on the other for its defining features, the other on the one for its continuing progress. In so doing, rationalization must propose at some level that a monolithic science impacts in a uniform way on society and culture. This, in turn, I have suggested, has fed through into an approach to the public understanding of science which, on the basis of similar assumptions, effectively defines any nonconformity to the purportedly uniform scientific orthodoxy as misunderstanding of one form or another.

The other strand of research into public understandings, however, points to a different way out of the dilemma. This approach seeks to observe how science appears in particular social contexts, focusing on how particular understandings—selective interpretations and interested formulations—are generated locally in relation to the local concerns and, to use Schutz's (1962) term, systems of relevancy of the actors involved. In essence, therefore, it abandons the assumption that science impacts in a uniform way on the social, treating science instead as the dependent variable in relation to the social. As such, this perspective points back to the sociology of knowledge and SSK in particular. From this perspective, science is no longer treated as having special standing, insulated from the social; instead, it is to be treated just like any other form of belief, as linked to and dependent on social processes and social context.

SSK focuses primarily on the detailed study of activities of scientists in practical settings such as the laboratory. It observes science in the making, seeking to open up the "black box" (Latour, 1987) that traditionally has tended to surround the actual activity of science, obscuring it from direct treatment by philosophical and sociological analysis. Thus, whereas traditional philosophical and sociological accounts tend to direct their gaze to the more or less abstract level of what makes science in general work, SSK focuses on the detailed particularities of what scientists actually do.

Out of this basic orientation, has arisen a number of positions varying on significant points of methodological and theoretical detail (see, e.g., Knorr-Cetina & Mulkay, 1983; Mulkay, 1981; Potter, 1996; Taylor, 1996). It is not my intention here, however, to address SSK closely, nor do I intend to attempt to summarize the substantial body of detailed studies of scientific practice, ranging as they do over all aspects of the history and current social practice and organization of the sciences, which SSK has stimulated. It is not necessary, for my purposes, that this be undertaken, especially when many valuable collections and overviews are already available (Ashmore, 1989; Barnes & Edge, 1982; Jasanoff, Markle, Petersen & Pinch, 1995; Knorr-Cetina & Mulkay, 1983; Lynch, 1993; Mulkay, 1979; Webster, 1991; Whitley, 1983; Woolgar, 1988). For the present argument, it is enough to consider the implications of the general approach of SSK for thinking about the public understanding of science (see also, Irwin, 1995). To this end, I want to return to one of the foundational statements of SSK, provided by Barnes (1974), where he also outlined to my knowledge, the first statement of the idea of science as a cultural resource.

The starting point of SSK, and Barnes' argument in particular, is the rejection of the traditional view of science as a special type of knowledge, in the particular sense of being somehow insulated—black-boxed—from external, especially social, influences on its development and progress. The contrast between the orthodox sociological treatment of science and religion is instructive here. The standard approach to religion in modern societies has always been to consider the social and/or psychological functions it serves—for example, as a form of legitimization of powerful social interests, or emotional compensation for the socially ostracized, or a means of normative integration, and so on. Whatever the specifics, the central point is that the belief is always interpreted in terms other than its own. Not so, however, science. Science, unlike religion, is allowed to be self-explanatory. The essential reason for this, Barnes argued, is that religious belief is viewed as fundamentally irrational, because false, whereas science is viewed as rational, because true, in the sense of corresponding to the reality of the natural world. As such, acceptance of same is taken to require no further explanation.

Given these assumptions, the task of the sociology of science was limited only to providing an account of the peculiar social conditions in which science is able to operate most effectively. This is how Merton viewed science, and his communicative model was designed accordingly. In a similar way, the task of the philosophy of science was confined to explaining the successful mode of working of science and thus to elaborate the significant grounds of demarcation between science and other beliefs (cf. Chalmers, 1982), the better to keep the house of science swept clean of impure influences, especially from whatever remaining irrationalisms might continue to infect the wider society.

Rejecting this, Barnes argued that, both the normative environment and the content of science is pervaded by the presence of the social and cultural. There is no a priori reason for excluding science from full-bodied socio-logical inquiry; it is a social activity and, as such, is amenable to study in the same terms as any other social activity. The black box can, and should, be opened up—hence, SSK.

However, Barnes appended a further level to the argument, addressing the role and position of science in the wider society, of greatest interest here. He adopted the Wittgensteinian view that the linguistic and normative context that surrounds and informs all human action provides a necessary, ever-present background against and within which any proffered explana-tion will be adjudged as meaningful as an explanation (cf. Antaki, 1994). More broadly, we arrive at the view that things called knowledge or beliefs can be treated as resources that may be drawn on by actors in order to make sense of situations for all practical purposes. It follows that those things that might be called beliefs can be studied for their pragmatic effects rather than their truth content.

A connection can be drawn with the recent radical rethinking, among social psychologists especially, of the nature of identity. Traditionally, identities have been thought of as relatively unified, singular, and stable qualities or attributions of people conceived as unified selves. Yet, under the influence of poststructuralist perspectives, there is now a growing interest in conceptions of people as having multiple identities, constructed out of the discursive resources available within postmodern culture (Mi-chael, 1996a; Shotter & Gergen, 1989). This entails a rethinking of the nature and meaning of beliefs along Barnesian lines: If beliefs were once considered as the more or less stable expression of attitudes by more or less stable human selves, with more or less stable identities attached to them, then this is no longer so. Rather, beliefs can now be thought of as discursive resources used for the construction of identities, attitudes, and so forth to be variably mobilized in relation to the pragmatics of particular situations and contexts. Thus, the type of belief expressed and the way it is expressed can be understood for the specific work it does in the specific interactional context where it appears (cf. Edwards & Potter, 1992).

It follows that statements of belief are likely to vary, even to the extent that apparently incompatible or contradictory beliefs may be expressed, with little or no apparent concern. This is because the baseline criterion of assessment of the adequacy of knowledge is not truth but normality, that is, what is considered normatively right, proper, and acceptable as a good—adequate and legitimate—explanation or understanding. Equally, however, a charge of incompatibility or contradiction may be used to attempt to undermine, invalidate, or delegitimize a given expression. This, then, may prompt further response addressed at attempting to rebut the charge in some way, either by questioning its legitimacy in turn, or by presenting a reformulation designed to repair the perceived inconsistency.

I have offered some elaboration from Barnes' basic point here, which begins to forge the connection I make between his position and DA. Barnes, himself, meanwhile, developed the argument to suggest ways in which social and cultural factors may become incorporated into science, through, for example, using metaphors drawn from everyday knowledge to depict and explain the novelties disclosed by scientific work. Of special importance here is Levi-Strauss' notion of *bricolage*. Barnes borrowed this, but not without first making the crucial transformation to suggest that it can be applied to science as well as to everyday sense-making: "As the *bricoleur* constructs devices for a wondrous range of purposes from a limited set of bits and pieces, so the primitive intellect constructs myth out of bits and pieces of culture. This delightful image ... is entirely appropriate to the scientist. He [sic], too, redeploys pieces of culture in new ways to perform new tasks." (Barnes, 1974, p. 58)

Equally, however, it can be suggested that nonscientists in modern society may draw on bits and pieces of modern culture, including science—although not necessarily exclusively so—in order to construct their myths, that is to say, their metaphors, images, conceptions, and other means of understanding and explaining the world.

It is helpful here to restate the contrast with rationalization. It must be recognized that the notion that science may be put to other uses within nonscientific contexts is perfectly acceptable to, and perhaps even anticipated by, the perspective of rationalization. However, because of the reification of science, its employment in this way entails an implicit judgement of correct and incorrect (or valid and invalid) use (and abuse). Science is treated, therefore, as something with a prior, clear, and identifiable meaning outside of contextual usage. Further, the expectation of rationalization is that other beliefs will, sooner or later, be replaced by this singular science, because of its greater truth value.

In contrast, the treatment of science as a cultural resource in Barnes' sense is diametrically different. Crucially, he said, "we must seek to *discover* it [science] as a segment of culture already defined by actors themselves." (Barnes, p. 100, emphasis original.)

That is, we should not as researchers, impose our understanding of science on actors—what we take it to be and to mean—but allow their meanings and understandings to come through.

This follows from the fact that, from the perspective of bricolage, science becomes merely one resource among others with which to construct understandings. As such, there is no reason to think that it will always be the same science employed, or the same bits and pieces extracted from it, nor that they will be used in the same way(s) or put to the same purpose(s). Further, these purposes may well be very far from those for which they were originally intended. Add to this, the internal complexity, fragmentation, and incoherence of science as Barnes viewed it—a view given empirical substance by SSK—then the possibilities of variation in the usage and adaptation of science for sense-making purposes in the wider culture multiply enormously and unpredictably.

This takes us into the highly debated question of the boundary of science. If science is extracted from its original context within the confines of the scientific community and put to other uses outside, then does it stop being science? If bits and pieces are drawn from science and adapted to suit other purposes not of their original intention, is this unscientific? Is it misunderstanding?

The problem with such questions is that they imply a sense of stability outside of social process. They wrongly presuppose that there is one boundary and one single, correct-for-all-time understanding. This is not so. The boundary of science must itself be seen as a matter of public debate and a moveable outcome of social processes (Barnes, Bloor, & Henry, 1996; Gieryn, 1983; Rothman, Glasner, & Adams, 1996; Taylor, 1996). As Barnes stated, the boundary between science and the rest of culture is a matter of actors' own understandings and demarcations and, as such, can be expected to show considerable variation between contexts.

For example, those demarcations propounded by philosophers, sociologists, and scientists themselves, become merely one set, employed by one particular group of actors. Even the authority attached to these actors and, therefore, the persuasiveness of the demarcations they present, can and should be treated as a social variable. Thus, rather than attempt to decide on the adequacy of these demarcations, it becomes of greater sociological interest to ask to what extent these (or similar) demarcations are found in other contexts and to what purposes they are used. What are the social practices of boundary demarcation used by actors, whether scientists or, ostensibly, nonscientists? It is this question that directs the analysis of creationism that follows.

Inevitably, however, this will raise the anxiety that such a stance is a recipe for complete collapse of the project of science. If science is not defined apart from nonscience, then does this not make all so-called scientific work, including presumably the present study, futile?

I am not convinced there is an altogether satisfactory response to such questions. I do not dispute that the relativistic implications of the sociology of knowledge and SSK especially can seem profoundly discomforting. Nor, however, do I think there is an acceptable alternative to them. Just as the samurai is able to appreciate the quality of workmanship of the sword by which he is impaled, so perhaps the relativist is one who, even as they are hoisted on it, attempts to study the qualities of their own petard![5]

This is the position of the discourse analyst (at least metaphorically). What needs to be stressed is that DA does not entail a commitment to ontological relativism, but to methodological relativism, or symmetry (Bloor, 1976; Potter, 1996). Methodological symmetry in the present study means that all sides of the story are treated equally regarding their possible truth content, precisely because questions of truth are not being asked of them so much as questions of how they formulate the propositions they deem to be true. It is not, then, a matter of trying to decide between these propositions in order to judge one view superior to the other(s) as a more accurate portrayal of reality. It may be the case that, in some sense beyond discursive representation, one version does provide a more accurate portrayal of reality than others, but that is not of interest. It is not of interest, because we do not have a sense beyond discourse; any and all senses are, by definition, articulated through, by and within discourse. What we are presented with is discourse, and it is this that we endeavor to study (cf. McPhail, 1996).

---

[5]In addition, it can be noted that the dilemma of the relativist shows the fundamental dilemma of science emerging here also. Relativism asserts particularism as a universal, thereby presupposing its own particularistic limits and anticipating its own demise (Woolgar, 1996). The dilemma is inescapable. The advantage of relativism is that it is aware of the irony in a way that the alternative is not. For further defense of relativism, see Ashmore, 1989; Edwards, Ashmore, & Potter, 1995; Woolgar, 1983.

What, then, does this mean for the status of the present study? It means two things. First, I deliberately suspend judgement as to the truth or otherwise of creationist knowledge claims. Unlike most other academic commentary on creationism, I do not set out to align myself with the orthodox evolutionary view endorsed by the scientific community. It does not follow from this, however, that I thereby endorse creationism. I am agnostic on matters of truth. Creationism is treated as (just another) discourse, to be studied like any other, for its discursive and rhetorical qualities. Although this is not an altogether new approach to creationist discourse (Prelli, 1989a; Taylor, 1992, 1996), it is still likely to strike some as unsatisfactory and perhaps even irresponsible. In this respect, it is notable that even the rhetoricians are careful to make clear that they ascribe no truth value to creationists' claims. I see this as a failure of rhetorical nerve and a breach of the principle of symmetry—a principle which, as Bloor (1976) insisted, must apply even to studies of science themselves, which I take to include this one. This brings in the second point.

What of the validity of this study? By what criteria is it to be judged, if not those of science? Such questions are misdirected. To adopt a position of methodological symmetry is not to claim somehow to be free of criteria of validation. DA adopts a qualitative methodology that presents interpretations of stretches of discourse extracted from a sample of material. As such, it can be assessed for the quality of its sample and for the adequacy of its interpretations, as well as for the validity of the inferences drawn from the substantive analysis. Where DA differs from the usual form of qualitative analyses is that, first, no claim is made to provide a definitive version (Gilbert & Mulkay, 1983; Mulkay, 1981) of the topic under study, and, second, following the lead of conversation analysis,[6] an effort is made to present sufficient quantity of the original text to enable readers to make their own interpretations against which to judge for themselves the adequacy of the analysts' own (Hutchby, 1996; Potter, 1996). This limits the amount of extracted material that can be analyzed in detail, thus, I have concentrated on covering those features of creationist discourse of most relevance to my overall concerns.

This is also a study of suasion techniques and, therefore, of rhetoric. To a degree, it is a contribution to the "rhetoric of inquiry" (Simons, 1989) or "rhetoric of science" (Campbell & Benson, 1996). There is an important shift in focus, however. Whereas the rhetoric of science movement is, as the

---

[6]For introductions to this, see Antaki, 1994; Atkinson & Heritage, 1984; Boden, 1994; Boden & Zimmerman, 1991; Drew & Heritage, 1992; Psathas, 1995; Sacks, 1995; Sacks, Schegloff, & Jefferson, 1974.

name suggests, concerned primarily with addressing the "rhetorical invention" (Prelli, 1989a) of scientists, understood in largely orthodox terms as members of the accepted institutional scientific community, the current interest, as should now be apparent, is in science outside this boundary. Nonetheless, although I am not trained as a rhetorician and do not adopt the traditional array of rhetorical terms (*topoi, stases*, etc.) nor seek to adjudge the quality of arguments, this is an attempt to analyze discursive processes of persuasion and as such may be assessed for its adequacy in these terms.

SCIENCE AS A CULTURAL RESOURCE, DISCOURSE ANALYSIS, AND CREATIONISM

Having presented, in general outline, my definition of science as a cultural resource—which is the usage of stuff drawn from science, especially by members of the nonscientific community, within everyday contexts, outside of those usually thought of as scientific—there remains a need to connect this general position with DA, in order to outline how the study of creationism will progress.

As previously noted, general support for viewing science as a cultural resource can be found in the recent research in public understanding of science, although it must be emphasized that this interpretation is not one put forward by any of the authors mentioned. To the best of my knowledge, there has been little development of Barnes' notion from when it was first presented. An exception is the short book by Cameron & Edge (1979) though, in my opinion, this perspective helps make sense of a range of other material.

If science is viewed as a cultural resource in this way, then a potential link with DA becomes apparent. Although specific applications of DA are discussed as they arise in the course of the analytical chapters, given the diversity of usage of the term *discourse* (Burman & Parker, 1993; Fairclough, 1992; Schiffrin, 1994; van Dijk, 1985), a brief outline of my perspective is appropriate.

DA refers to the perspective developed principally by Potter & Wetherell (1987, 1994; Wetherell & Potter, 1988, 1992). It focuses on the ways in which members of society formulate their accounts of their own and others' beliefs and actions (cf. Gilbert & Mulkay, 1983, 1984). This shows the influence of one of the main sources of the perspective, ethnomethodology (Garfinkel, 1967; Turner, 1974). Like ethnomethodology, DA is interested in accounting practices—the explanatory narratives of their own and others' actions that people construct—and pays close attention to the particular details of accounts given in different contexts. Detail is considered important, because it is in the detail that differences become apparent, and it is in the differences that people mark out what is currently being considered as worthy of consideration, and in what precise way(s) they want it to be considered. There is, therefore, an emphasis on variability within DA—variability in the way words and terms are used, and in the intended and derived meanings that are given to and taken from them. Thus, DA assumes words are likely to be unstable things that will shift in meaning and semantic texture, depending on who is talking to whom and under what circumstances. In this, there is a link with the rhetorical position developed especially by Billig (Billig, 1987, 1991; Potter, 1996; Potter & Wetherell, 1994).

Following Billig et al. (1988), the basic proposition in this book is that modern society is marked by a culture of arguments and dilemmas; these arguments and dilemmas are to be found in distilled form in all kinds of words and phrases commonly used in everyday parlance—commonplaces. Further, I propose that precisely because of the heterogeneity (Law, 1991) of potential meaning of words—their ambivalence potential (Fairclough, 1992)—this may give rise to efforts at synthesis, to resolve the (apparent) contrasts and thereby generate new meanings and novel arguments, for which stability and closure are claimed. That such claims may be futile—for reasons both internal to the proposed resolution, in which points of tension and ambiguity will remain, and external, as any proposed resolution may always be resisted—does not stop them being made (to the extent, indeed, that efforts may be made to enforce them by means more tangible than the symbolic weapons of argument).

Creation science is to be understood in these terms. It is an attempt to resolve the (supposed) opposition between science and Christian belief through the construction of particular accounts of each drawing on the argumentative and rhetorical resources they provide and exploiting the dilemmatics within and between them. The outcome is a position in which the original contrast between science and religion is dissolved and, in fact, said itself to be an outcome of the erroneous perspective (modern, evolutionary science). This (claimed) dissolution is achieved discursively through the construction of meanings—the meaning of the world (empirical evidence) and the meaning of the Word (the Bible)—to produce compatibility. I call this process *discursive syncretism* and it is discussed in detail in chapter 5.

This shows, then, that meanings are not stable; they are multiple and their multiplicity provides the basis for disputation. Further, and most importantly, this disputation is not merely academic, in any of the senses of the word. Disputation is central to everyday life. Even if not overtly occurring, disputation is implicit, in the sense that all situations may be thought of as the outcome of other disputations (and the attempts to stabilize them), and everything that is said in a given situation bears the mark of the battles that have been fought in the past (Volosinov, 1973). Thus, words are weapons, and arguments are shields in all contexts—although clearly the specific words and arguments that arise or have the potential to arise will vary with the specific context.

Nonetheless, as a general rule, instability and heterogeneity persist. One of the main reasons for this is precisely because the meaning of the situations that confront us is itself often unclear. Situations often present us, not with one way of understanding and dealing with them, but with dilemmas, that is, with apparently contrasting ways of understanding and dealing with them, either side of which may seem equally favorable or unfavorable. Putting this another way, we are able to characterize situations as either one way or the other, by drawing on the commonsensical resources available to us. These resources can be said to come from general ideologies, which together make up the commonsense lifeworld (Berger & Luckmann, 1971; Habermas, 1984; Schutz, 1962) of the society. Ultimately, the dilemmas we confront exist because of the internal fractures to be found within—and between—these ideologies (cf. Billig et al., 1988).

Thus, this book adopts the now widely accepted view that any given situation is likely to be open to more than one way of characterizing it, not just by different people, but by the same individual. In consequence, whatever way of characterization is eventually adopted runs the risk of being confronted by an (at least one, possibly more) alternative characterization—and, in a sense, may actually encourage the presentation of an alternative—which, in turn, may give rise to a process of argumentation in which participants make accusations about and seek justifications for the views they have taken. In consequence, any given characterization contains an "evaluative accentuation" (Volosinov, 1973, p. 22), or a rhetorical charge, in relation to the potential alternatives and the argumentations implied.

This, then, is my approach to the texts of creation scientists and the texts of sociologists commentating on creation science. I seek particularly the ways in which they characterize science, the public understanding of science, and creation science itself. This is not, therefore, an attempt at an exhaustive analysis of creationist discourse. It is a directed study, which should be seen as contributing to the discourse–rhetorical analysis of science in modernity.

As should also be clear from the previous outline, when I use the term *rhetoric,* it is not with any pejorative intent. From the Billigian perspective, rhetoric takes on much more of its classical sense, referring to forms, techniques, and procedures of persuasion (Billig, 1987; see also Prelli, 1989a; Simons, 1989). This sense of the word penetrates deeper into the structure of discourse than does the more typical modern sense of rhetoric, referring merely to superficial stylistic flourishes, or deliberate obfuscation in speech or writing.

Most discourse analyses of science to date have focused directly on scientists, that is, members of the orthodox scientific community (Gilbert & Mulkay, 1983, 1984; McKinlay & Potter, 1987; Mulkay, 1985; Mulkay & Gilbert, 1982; Mulkay, Potter, & Yearley, 1983; Myers, 1990; Potter, 1984, 1987, 1988), although there are now some studies in this vein of science in more public settings (Ashmore, Mulkay, & Pinch, 1989; Michael, 1992, 1996b; see also Hess, 1993). However, there has been little attempt so far to address the broader implications of DA for the understanding of science in modernity, in relation to traditional modes of sociological theorizing about this. With the burgeoning interest in the public understanding of science over the past decade, it is relevant that the DA perspective should have some proper representation in the debate, as it offers a quite different approach to that of traditional sociology, with radical implications. Here, especially, the connection with the conception of science as a cultural resource is persuasive. DA also speaks in terms of resources, although in a manner focused on a specific level of inquiry. Despite the similar terminology, then, it is not immediately obvious how speaking in terms of discursive resources might relate to the cultural resources provided by science. It is possible, though, to suggest some lines of connection drawing from the existing corpus of DA work. In particular, the focus on interpretive repertoires and argumentation is especially provocative in this respect.

A key feature of DA is the identification of interpretive, or linguistic, repertoires, that is, stock modes and techniques of expression, which display systematic variation and contrast (often, although not always, varying with context). The term was used by Gilbert & Mulkay (1983, 1984) to refer to the different voices of expression used by scientists when discussing their own and other scientists' beliefs and actions. On some occasions and in some contexts, they tended to adopt an impersonal mode of expression—the empiricist repertoire—at other times, a more personalized mode—the contingent repertoire. Recent research showed these and other repertoires being used in other contexts (Edley, 1993; Marshall & Raabe, 1993; McKinlay & Potter, 1987; Potter & Wetherell, 1989; Wetherell & Potter, 1988, 1992; see also, Macnaghten, 1993; Wooffitt, 1992).

In a general sense, interpretive repertoires can be thought of as providing sets of argumentative or rhetorical resources. As analyzed by Gilbert & Mulkay (1984; cf. Mulkay, 1985), for example, it seems apparent that what they call repertoires consist of various representational resources by which scientists may justify and legitimize their own actions and beliefs, concurrently undermining and delegitimizing those of opposing scientists (cf. Potter & Wetherell, 1987). They also provide ways of accounting for their opponents' (apparent) errors (Mulkay & Gilbert, 1982). Thus, it is only a short step to a more general focus on forms and modes of argumentation, or rhetoric (Potter, 1996). This type of focus seems especially appropriate when considering groups involved with some kind of debate with scientific orthodoxy, as creationists clearly are.

As it happens, a key feature of arguments about and involving science is a concern with the nature of science itself. This is true, of course, within the scientific—and, more broadly academic—community; however, it is also true, as I have argued consistently, in the nonscientific community. This is also, perhaps, a more immediately obvious sense that one might take from the term *science* as a cultural resource. If one is interested in what science means to people, then it seems a useful starting point to look for their way(s) of characterizing science. What model(s) of science do they use and how—argumentatively—do they use them?

This is also an issue of particular pertinence with respect to creationism, in that sociologists discussing creationism often make reference to the model(s) of science that creationists use. This is usually done, however, in order to claim that creationists *misunderstand* science. Thus, this would seem to be an especially interesting and appropriate feature to focus on as a way of establishing precisely what creationists' understanding(s) of science is (are) and to what extent it is helpful to consider them, less for the misunderstanding they display, and more for the rhetorical work they do within their context of presentation. This helps to expose clearly the distinction between traditional sociological approaches, drawing on rationalization, and the rhetorical approach of DA.

Thus, these considerations define the structure of the remainder of this book:

Chapter 2 introduces creation science more fully, through a discussion of its representation in sociological texts, offering explanations in terms of rationalization. This serves a dual aim. It provides a more focused illustration of the argument in this chapter to show that the discourse of rationalization reproduces the dilemma of science despite itself. This argument is developed by showing that what is called creation science in these texts is constructed in relation to other constructions: of science, and of the *public understanding of science.* Although there is a mutually reinforcing rhetorical interplay between these constructions, it is one that produces variations in the accounts of creationism offered, variations that display the tensions within the discourse of rationalization.

Chapters 3 to 5 comprise the bulk of the book, collectively providing the analysis of creationist discourse, using pamphlets published by the British Creation Science Movement (CSM). The analysis focuses in turn on three things: their usage of different models of science; their use of the empiricist and contingent repertoires; and their reading of the Bible through what I call a *fundamentalist repertoire.* The results show that creationists are involved in the management of three interwoven problems. The first is similar to that confronting scientists in dispute with each other: Creationists must account for the existence of an alternative version of reality (the world), in the form of evolution, when they insist that their creationist version is not only correct, but plainly so.

The second problem arises because, not only are creationists in dispute with orthodox scientists but also with nonfundamentalist Christians, who offer versions of the Bible (the Word) alternative to creationists' own. Thus, creationists must show that their version of the Bible is the correct one and these other versions mistaken; to accomplish this they use the resources of the fundamentalist repertoire. However, this is further complicated by the third problem: They need to ensure their version of the world conforms with their version of the Word; that reality and the Bible say the same things. This involves them in a dual process of reading, interpreting science through the Bible, and interpreting the Bible through science at the same time. I call this dual process of reading *discursive syncretism,* and it is analyzed in chapter 5.

Chapter 6 then addresses rationalization again. Here, I return attention to sociological texts, offering analyses of the role and position of science in modernity from the framework of rationalization. Specifically, I consider selected writings by Habermas, Bell, and Lyotard. Although this is by no means an exhaustive coverage of recent theorists of science in modernity, it does provide representation of a range of different characterizations of the current situation of advanced industrial societies, as late capitalist, postindustrial, and postmodern. Equally as important from the present point of view is that, despite these differences, the writings of these theorists all display the central features of rationalization discourse, which gives rise to various ambiguities and tensions in their accounts of science and the public understanding of science. The struggle with these ambiguities and tensions leads to the introduction of communicative versions of science, that do not concur with the rhetoric of rationalization. Further, they reintroduce the central dilemma of science in modernity, between universal and local forms of knowledge.

In chapter 7, I attempt to pull the strands of the analysis together in an argument developing the idea that there is a shared discourse of science in modernity constituted from the dilemma of science, with its dual characterization as both a universal form of knowledge and a local product of particular human actions and beliefs. This dilemmatic discourse is articulated within the lifeworld of modernity, whether within the academy, in the tensions between and within rationalization and the sociology of knowledge, or within the wider culture in the use of science as a cultural resource. Thus, sociologists and creationists draw upon science in comparable ways using similar rhetorics, albeit to contrasting ends. Sociologists seek to construct accounts of a universally valid science that will also explain anomalies like creationism. Creationists seek to construct accounts of science and the Bible that can be presented as compatible, meanwhile explaining the existence of anomalies such as evolutionism and nonfundamentalist Christianity. Further, the use of Christianity as a resource shows that tensions exist not only within discourses (e.g., science and religion), but also between them. From this, I develop an argument in defense of a view of modern culture as a culture of agonistics, or competing *logoi*, and call for further research into the dilemmatics of science and other discourses within modernity.

2

Creating "Creation Science"

In chapter 1, I argued that many sociological accounts of the role and position of science in modernity adopt the discourse of rationalization in which a monolithic science is presented as impacting in a uniform way on the wider culture and consciousness of the modern. It is now time to focus on how this discourse works in those specific accounts directed at understanding and explaining creation science. This involves showing how creation science is constructed to suit the rationalization hypothesis. Within this hypothesis, science is presented as a separate kind of thing, which behaves in one kind of way in relation to the wider society: Science acts and society reacts. Given this, creation science itself is presented as a response to this influence. It is also presented as not science, as a thing outside of science, clearly demarcated from it, and as a phenomenon that has reappeared in an unexpected and surprising way, like an unwanted weed, thrusting back into a tidy social order when it was thought to have been eradicated. A key point is that creationism is constructed as new, albeit in the sense of a return of the departed, and like science only in the sense of being another monolithic entity, albeit one that is essentially religious, rather than scientific.

In contrast to this story of creationism, however, it is possible to construct an alternative in which it appears as neither new, nor monolithic. Seen this way, it is not creation science that is the alien entity, but evolutionism, which appears as an external interfering presence imposed by imperialistic powers.

To further add to the curiosity, both of these stories of creationism—as both new/not-new and monolithic/not-monolithic—appear in the accounts which attempt to explain it in terms of rationalization. My argument is that this is inevitable and demonstrates the inadequacy of the rationalization hypothesis. Rationalization seeks to impose a uniform view of such phenomena, but multiplicity keeps emerging, despite best efforts to contain it.

This itself shows up the inadequacy of the discourse of rationalization and points toward the alternative idea of science as a cultural resource. This chapter concentrates on the question of newness; in chapter 3, I examine the internal variability of creationism more closely.

CREATING "CREATION SCIENCE"

First, however, the term *creation science* needs some clarification. I have repeatedly referred to creation science, but have yet to present any kind of systematic description or definition of what it is. Unfortunately, this is not as straightforward a task as it might seem, precisely because of the emphasis placed by discourse analysis (DA) on multiplicity and variability (Potter & Wetherell, 1994). It is this that sets DA apart from the more orthodox sociological approaches that are the subject of criticism here. From these perspectives, it is accepted that a simple and coherent statement of subject matter is possible. For the discourse analyst, however, this merely raises a dilemma. Any such account is a construction. As such, it serves pragmatic purposes within the context in which it appears. It is not just a description, but a social intervention.

If, for example, it were to be proposed, for the sake of illustration, that this were not a DA study, then it would be fully in keeping with established expectations now to provide an outline of creationist beliefs, building on the basic definition given in chapter 1, that creation scientists are Protestant fundamentalists who reject the theory of evolution in favor of the book of Genesis account of origins and do so for what they claim to be scientific reasons. Before this was done, of course, it would also be necessary to ensure that a suitable disclaimer was included to the effect that this summary of beliefs should not be taken as reflecting the author's own views, but had been culled from creationists' own writings. Thus, it would be emphasized that the summary should not be read as an endorsement of the claims made by creationists, and further that it is given in full awareness that these are hotly disputed by defenders of evolutionary theory. The point would also be made, however, that it is not the purpose of this study to consider these defenders' arguments. Rather, this study is concerned with what significance creationism might hold for understanding the role and position of science in modernity; thus, it is less concerned with the debate with evolution as such, and more with the use of science within creationists' own writings.

Having so distanced himself (he is male) from his own summary and, indeed, rejected ownership of it, the author might then go on to detail creationists' beliefs. This might consist of a series of statements based on his reading of creationists' texts, which would be presented as a representative summary of their essential content, thereby constituting what might be called a *definitive version* (Gilbert & Mulkay, 1983) of creationism. Such a version would be likely to consist of some or all of a range of commonly identified features concerning both their criticisms of evolution and their claims in support of creation. Among these might be included some or all from the following list:

1. Evolutionary theory is speculative and unscientific;
2. Evolution contradicts the Second Law of Thermodynamics, one of the foundation stones of modern physics and chemistry, which states that the overall level of disorder in a closed system always increases with a change of state, whereas evolutionary change seems to produce increasing order and complexity;
3. Evolution ignores falsifying evidence, such as the lack of transitional forms between species in the fossil record (e.g., from fish to reptile, or reptile to mammal, or ape to human) and the harmful consequences of the majority of mutations (e.g., sickle-cell anaemia);
4. There are many anomalies that do not fit evolution, such as the discovery of fossilized remains of creatures in rock strata of ages earlier than those in which the creatures were supposed to have lived, and also living fossils, creatures that were thought to have died out eons ago, but have been found alive and apparently unchanged from their fossilized forms, despite the supposed millions of years in between;
5. Such fossil evidence does not support evolutionary development of species, but rather the creationist notion of fixed, unchanging kinds, as created by God, according to the Bible;
6. Evolution also ignores evidences against a billions-of-years old universe and that support a much younger age, of the order of thousands of years, such as the relatively thin layer of dust on the moon's surface and the relatively thin layer of sediment on the bottom of the earth's seas;
7. The uniformitarian geology of evolution also ignores the evidence for catastrophism in such things as the overthrusting of rocks, the formation of coal beds, the distribution of fossils, the destruction of the dinosaurs, and the freezing of mammoths in glaciers; these things could not be caused by gradual processes of change spread over millions of years, but only by some catastrophic event or events, causing a sudden, widespread transformation in the earth's geology and climate, such as might be brought about by a worldwide flood, as recorded in Genesis;
8. Only creation can explain such things as the Cambrian explosion (the apparent sudden eruption of complex life-forms in the Cambrian rock strata),

the numerous evidences of design in the structure and organization of the universe, the solar system, and the ecological and biological systems on the earth, and the existence of moral beings that know purpose—humans;

9. Similarly, only creation can explain the beginnings of the fundamental process of life in the production and reproduction of genetic material, the complexity of which is unimaginable without a directing intelligence;

10. Archaeological and demographic evidence supports the view that humans have not, and cannot have, existed on the planet for more than a few thousand years, and that they have expanded outwards from a point centering on or near Mount Sinai.[1]

The trouble with any such summary list of points, however, is twofold. First, it can always be added to, if only because at the present time, new writings by creationists are continuously being produced, and not all of these writings are merely restatements of already existing views. In any case, there is always more detail to such expressed beliefs, which can never be fully captured except by full reproduction of the original text(s). This may seem obvious, but it is not trivial, especially not for the discourse analyst, because what is being attended to is the pragmatic work undertaken in each new statement or formulation.

This brings up the second point. What is critical about a statement is not merely what is said, but also the way it is said; indeed, in an important sense, these two things are inseparable, as the way something is formulated tells us important things about its meaning. For example, in my own text previously, I deliberately referred to myself as the author, in order to convey something of the artificiality of the notion that an author can distance themselves from what they write. Referring to oneself in the third person, or as an anonymous author is one established technique that may be used to try to achieve such distancing. I chose this rather clumsy and unconvincing method in order to display this plainly, something that I could not do as well by continuing to speak in the first person.

The broad point behind this, then, is that statements are always interventions, no matter how well made up they may be to appear otherwise. Thus, any attempt to define creation science, such as by providing some list of creationists' beliefs, will necessarily be a construction tailored to suit specific, occasioned requirements in the context in which it appears. I attempt to show this is the case through a brief examination of some representations of creationism in the existing academic literature. I begin

---

[1]This list of claims is drawn principally from recurring features of CSM pamphlets, but I have also used Cooper (1995), Morris (1974), Rosevear (1991), and Whitcomb & Morris (1969). For criticism of these and other creationist claims, and defenses of evolution against them, see Godfrey (1983), Kitcher (1982), Montagu (1984), Ruse (1982), and Wilson (1996).

with some consideration of efforts to locate creationism within the broad territory of Christianity as a whole.

FUNDAMENTALLY SPEAKING

In my earlier bald definition of creationism, I asserted that creation scientists are Protestant fundamentalists. This is, at best, only partly true. First, and most obviously, not all creation scientists are Protestants. There are Catholic creationists (Miller, 1984; Turner, 1982) and non-Christian believers in at least some theory of creation, who may also define themselves as scientists, such as Dr. Chandra Wickramasinghe, who took the stand on behalf of creationism in Arkansas in 1981 (LaFollette, 1983; Overton, 1982; Wilson, 1996).[2]

Further, there are numerous divisions among the Protestants themselves, prompting some efforts at categorization. In Britain, Barker (1980) proposed an "ideal–typical characterization of ... scientists' positions vis à vis Christian religious dogma," within which creationists are defined as fundamentalists, in contrast to other groups, such as "orthodoxy (evangelicals)" and "liberal[s]" (p. 284). Even among only the creationists, however, Barker distinguished the Evolutionary Protest Movement (EPM; now the Creation Science Movement or CSM) from the Newton Scientific Association (NSA) on the grounds that the former "are very careful not to resort to the Bible but to stick rigidly to secular references in their public statements" (Barker, 1980, p. 285). EPM/CSM, on the other hand, are more eclectic.

In the United States, where creation science figures more prominently in public arenas, these types of distinction have been accorded considerable significance by some analysts. Toumey (1991), for example, presented an account of the internal divisions among fundamentalists in respect of their "positions on the moral authority of science." (p. 693) He distinguished eight such positions, ranging from the "extreme hostility to science" (p. 694) displayed by Jehovah's Witnesses, through various ambivalent and qualified positions. At the other end of the scale are those, such as the Foundation for Thought and Ethics, who, like the NSA, are careful to "omit explicit mention of God or religion" (p. 697), and the American Scientific Affiliation (ASA), who are said to be caught in an "evangelical dilemma" of sharing religious values with other creationists, but concerned not "to

---

[2]The term *creationism* may also, of course, be applied to many other, non-Christian beliefs, although it is perhaps less likely that any scientific status is claimed for them. See Barker, 1996; Layton, 1996.

jeopardize its scientific credibility" by coming out, as it were. Toumey (1991) used these distinctions to argue that they demonstrated that "the social role of scientific authority is both varied and unstable" (p. 697).

This view agrees with the notion of science as a cultural resource outlined in chapter 1. However, it is less clear that Toumey's attempt to distinguish creationist positions goes far enough in bringing out just how varied and unstable is the authority of science. It is notable that, in reference to the eight positions and their representatives, he observed, "the views of some of these groups overlap, and ... numerous additional positions ... can be added to this scale by locating ideological niches between and even within the[m]." (Toumey, 1991, p. 695)

This, I would suggest, is a more significant point than Toumey made out, for it seems to call into doubt his categorization exercise. It is unclear what advantage is gained from the construction of such ideal–typical positions, if it is already apparent that they do not capture the actual diversity within and between stances.

I prefer to view Toumey's categories in the following way: In establishing this range of positions, he identified a number of argumentative stances, which he then equated with specific creationist groups. However, this equation discounts the movement between positions that arises from the capacity of members to adopt these stances flexibly, such that members of specific groups may, at times, adopt argumentative stances equated by the analyst with other groups altogether. It is better, then, to view these argumentative stances as commonsense resources potentially available to all the various creationist groups. This is not to deny that some groups (and, perhaps, some individuals within each group) may tend to adopt one stance more commonly and regularly than others, or may never be found adopting some stances. However, in recognizing that such positions are potentially available as rhetorical characterizations, we refine our analytical approach and open up a more flexible conceptualization of the social role of science.

To illustrate. In discussing the "extreme rejection of science" position, Toumey (1991) referred to the Worldwide Church of God, stating that its "views on creation and evolution are almost identical to those of the [Jehovah's] Witnesses." He quoted from the writings of the founder of the group, Herbert Armstrong, as follows:

"Has either science or education *proved* the evolutionary theory? Emphatically they have not! ... Has either science or education *proved* the *non*-existence of a personal Supreme God? ... Emphatically they have not! Why, then, do so many great minds who profess knowledge and wisdom doubt or deny

God's existence? Simply because of something inherent in human nature—something of which they are ignorant—a spirit of vanity, coupled with
hostility and rebellion against their Maker and His authority!" (Armstrong,
quoted in Toumey, 1991, pp. 685–686; elisions and emphases as in source.)

Very similar words can be found in the publications of the British CSM,
such as the following example[3]:

The concept of evolution has not been scientifically established, in spite of
over a century of intense scientific investigation .... This should surely be
ample evidence that it cannot be claimed to be true science .... That evolution
has taken place somewhere, somehow and somewhen is a faith in the
unknown. It is the result of the firm rejection of the evidence for divine creation
in an endeavour to deny the existence of the Almighty God, and the authority
of His Word. 'The fool has said in his heart, there is no God.' (Ps 14, 1)
(Pamphlet 277, p. 4; my elisions.)

However, despite this apparent similarity, many other quotations could
be drawn from my sample of CSM's publications, which present views that
seem more in keeping with other of Toumey's positions. This is not
surprising, because CSM often draws on material published by such American creationist organizations as the Institute for Creation Research (ICR)
and the Creation Research Society (CRS). These groups were categorized
by Toumey (1991) separately from the Worldwide Church of God, in the
position he called *scientific creationism*. This is said to be "the form of
creationism most concerned about idioms of scientific respectability" and
that "resonates very closely with the widespread respect for the moral
authority of science ... that flourishes in American culture." (p. 694) Quite
different, it would seem, from the extreme hostility of the Witnesses.

It may be, of course, that there are some substantial differences between
British and American creationists that would account for CSM's apparent
eclecticism (Locke, 1996). Nonetheless, I think it preferable to see in
Toumey's positions some aspects of the range of argumentative stances that
may be adopted, selectively and variably, by creationist groups. In attempting to pin Christian fundamentalists to particular positions, Toumey effectively reifies the positions themselves. Although this may suit some
analytical purposes, it imposes artificial limitations on our grasp of creationist discourse. To explore this fully, there are advantages in concentrating

---

[3]In all extracts taken from CSM pamphlets, I have endeavored to remain consistent with the
punctuation, spelling, emphases, and other textual features as found in the original source. References
are given by pamphlet number, rather than author. A separate list of pamphlets by number is appended
to the bibliography.

on just one group of creationists, as this enables the full range of their arguments to be covered in detail. This is the strategy of the present study, focusing on CSM, the flexibility of whose discourse is a major feature of the analysis developed in later chapters.[4]

If it fits anywhere into Toumey's scheme, CSM would seem to go into the category scientific creationism—although, arguably, this is only because CSM draws some of their own material from the American creationist groups Toumey placed here (ICR and CRS). They also publish their own material, however, largely in the form of regular bimonthly pamphlets (more details of which are given in chapter 3). In the rest of the discussion of creationism that follows, then, any creationist material referenced is taken from these publications. Thus, for present purposes, it is the content of these publications that constitutes creation science.

EXPLAINING CREATION SCIENCE

In ensuing chapters, the content of these publications is discussed in considerable detail. For now, though, let me return to the question of how creationism has been understood by orthodox sociology. I have already given some indication of how creationism is constructed as a category to suit analytical purposes. Now I show how this is accomplished to fit in with the rationalization hypothesis.

For the most part, sociological studies of creationism focus on the groups mentioned in Toumey's category scientific creationism (ICR and CRS, and CSM in Britain) and one other major group, the Creation-Science Research Center (CSRC), categorized by Toumey (1991) as "litigation-contingent creationism." Although other United States groups are often mentioned—such as the Bible-Science Association and the ASA—these three have attracted most interest. This would seem to be a reflection of their activity over recent decades, especially CSRC's involvement with the claim for equal time in the teaching of creationism in United States state schools (Nelkin, 1982; Toumey, 1994). It also seems to be due to their having all been connected at some point with the figure of Henry M. Morris. Morris

---

[4]In the British case, in fact, the focus on one group is almost mandatory—at least it has been for me! The number of active British creationist groups is much smaller than in the United States. In practical terms, I have found there is only CSM, as, although other groups are sometimes mentioned—including the NSA (if they still exist), the Bible Creation Society, and a Catholic creationist group—I have been unsuccessful in contacting them. Other researchers, however, seem to have had more luck in this respect (Coleman & Carlin, 1996).

is often seen as the key figure in stimulating what is widely presented as a resurgence of creation science since the 1950s (Kitcher, 1982; Livingstone, 1987; Nelkin, 1982; Numbers, 1982, 1987; Shapiro, 1986).

If there is a standard view of creationism in the sociological literature, it is one that sees it as having in some way reappeared on the stage of modernity at this time, having apparently not been heard of and essentially forgotten since the 1920s. This representation is central to the manner of its accounting. I first set this out as a general description, before considering specific arguments in more detail.

Thus, in the United States case, the rise of creationism is linked to the American response to the perceived threat posed by the Russian Sputnik program. Fearing the greater advance of Soviet science and technology, a Federal program to overhaul the teaching of science was established, which included incorporating the theory of evolution into biology texts (Barker, 1985; Dolphin, 1996; Nelkin, 1982). It is this which is seen as the proximate cause of the resurgence of antievolution sentiment among parents in local communities, beginning in California (Nelkin, 1992a; Toumey, 1994) but quickly spreading to a range of southern and mid-Western states. A succession of legal battles has occurred since then, in which state educational authorities have variously considered or sought to legislate the teaching of creation alongside evolution, in response to creationist pressure, in some cases leading to Supreme Court action (Gerlovich & Weinberg, 1996; Katz, 1996; LaFollette, 1983; Schneider, 1995; Shermer, 1991; Toumey, 1994).

Creationists have continued to be successful, however, in shaping the content of school textbooks to resist the presentation of evolution as a fact, or even as an undisputed theory, and in making their own texts on creationism available as alternatives. Indeed, they have been so successful in establishing their organizational base that they now have their own universities, as well as research centers like the ICR. Although there appears to have been some slowdown in activity since their failure at Arkansas in 1981, United States opinion poll evidence shows continuing widespread public sympathy for their cause (Taylor, 1992; Toumey 1994).

However, although the specific circumstances of Cold War paranoia are given as providing the historical moment for resurgence, there is a more general set of conditions that is also called upon to account for creationism. These conditions are taken to apply to Britain as much as the United States. They include the growth of Big Science, already a tendency of state and business bureaucratization before the Second World War, but further consolidated as one outcome of same; the growth of antiscience attitudes, itself in part a response among sectors of the public to Big Science (and especially the atomic events that came to symbolize it); and a more general disaffection associated with the loss of moral and communal stability in advanced industrial societies, added to which, the increasing specialization of the expert opens an ever-widening gulf between science and the laity.

These factors amount to an increase in the sense of alienation from and suspicion toward corporate representatives, among which science and technology are increasingly counted, and a growing intolerance of what is perceived as state interference in local matters. Thus, creationist resistance to evolution, although seen in its specifics as very much a reflection of local environments previously insulated from the influence of the secularized world, is nonetheless aligned with the more general disquiet of a society cast adrift in a sea of existential angst, without benefit of either the moral rudder once provided by now outmoded traditional beliefs, nor guidance from an increasingly distant scientific community. The old priests are without credibility, but the new ones have, like monks, turned their eyes from the world outside in favor of the visionary experience found in the cloisters of the high tech lab.

Creation science, then, is the bastard product of a disenchanted world; God remade in a test tube, a perverse issue from the unsatisfied desires of traditional morality in desperate coupling with an unbounded cognitivism that does not know when to stop. It is, in other words, as Gilkey called it, a "deviant," "bizarre" syncretism (Gilkey, 1987).

However, on closer inspection, this account of creationism begins to look a little more suspect. There are three aspects in particular that when scrutinized, lead to serious doubts. They concern science, the public understanding of science, and creation science itself, respectively.

In the case of science, central to these accounts is its presentation as a monolithic entity, with clear boundaries demarcating it from other forms of knowledge or areas of culture. This applies, regardless of whether science is characterized in conventional or more straightforwardly empiricist terms. Nelkin (1982), for example, presented a clear conventionalist account of science, drawn from Merton and Popper. Thus, Nelkin said: "Scientists accept theories and teach them, not because they represent 'truth,' but because they are accepted by the scientific community as useful explanations of reality." (p. 186) Similarly, she said, "most scientists understand their ... work [as] approximate, conditional, and open to critical scrutiny." (p. 189).

This characterization is central to Nelkin's view that the public understanding of science, and of creationism as an aspect of this, is based on a misunderstanding of what science is really about. In particular, there is a failure to recognize the moral and normative limits of science: "By seeking external political approval of the validity and justice of their arguments, they [creationists] ignore the constraints imposed by the norms of the scientific community" (Nelkin, 1982, p. 157). As Prelli (1989a, 1989b) showed in some detail, this kind of argument was used very effectively by members of the scientific community, at the Supreme Court hearing in Arkansas, to exclude creationism from their ranks. Given Nelkin's (1982) involvement for the plaintiffs in this trial, it is perhaps unsurprising that she should seek to incorporate this characterization of science into her socio-logical account. Yet, there is something odd about it, because, as she herself noted, scientists also frequently break such norms when speaking in the public sphere. Scientists, she pointed out:

> respond to criticism with their own kind of fundamentalism, emphasizing the neutrality and apolitical character of science and the weight of evidence that supports scientific authority.

> However, in the heat of public debate, this neutrality breaks down .... Biologists and creationists alike ... bemoan the moral, political, and legal implications of the alternative ideology. (p. 191)

The question then must surely be, if the real scientists behave in this way, then what exactly is there that enables us to distinguish them from the fakes like creationists? Similarly, we might also ask what it is about public debate that should make it somehow more heated than the debate that takes place inside the laboratory?

This familiar rhetoric is part of the contrast that is constructed between science set apart in its own protected realm, and science as it appears exposed to general public life. The first is supposed to be different from—and more trustworthy than—the second. The crucial point, though, is that this is a rhetoric (Prelli, 1989a, 1989b). It is, as Gieryn, Bevins, & Zehr (1985) showed, doing boundarywork for science, in their view func-tioning to maintain professional scientists' monopoly over the market for scientific knowledge (p. 403). However this may be, what is of most concern to the present perspective is that the rhetorical nature of this proposed demarcation becomes clear from features internal to Nelkin's own account, in the form of the intruding presence of an alternative view of science that breaks down the contrast between science and the public. What can be seen, in fact, is the emergence into Nelkin's account of the same dilemma discussed in chapter 1, between science as an asocial form of knowledge, standing outside of society, and science as a product of human, social actions. The attempt to understand science in conventional terms, invoking a special set of norms, is one effort to resolve this dilemma. The projected resolution breaks down though when the public activities of scientists are considered.

To put it another way, these public activities, in which scientists themselves are said to be acting in ways that contravene the norms of the community, have to be characterized as a contravention, in order to save the initial model that purports to describe the norms contravened. The model itself is not questioned, because to do so would place in jeopardy the demand to insulate science—*real* science—from those things that are held to be not real science—things like creation science. The upshot is that, in effect, creation science comes to be defined by the model: It has to be made a certain way in order to fit; it is constructed to suit.

A similar point applies in the attempt to account for the resurgence of creationism by linking it to so-called antiscience movements. This can be seen from Barker's (1985) essay. The basic thrust of the argument here is broadly compatible with Gellner's analysis seen in chapter 1. In this argument, the growing distance between the scientific specialist and the layperson, coupled with a more general moral malaise in modernity, has produced a backlash reaction against what is presented as the widespread prewar faith in science and technology—the "[h]opes that science was to become the God of the future," as Barker (1985) stated. This in turn fuelled: "a growing desire for order and certainty ... to reaffirm one's roots. In some quarters, this was translated into a resurgence of traditional belief." (p. 197) Hence, creationism, with its overt Christian fundamentalism.

Why though, if this is part of a general rejection of science and technology, should it take the form of creation science? Barker's (1985) argument was that, although its ethics may be questioned, science remains unassailed as the epistemological authority, the sole source of valid knowledge, with the consequence that: "If, in the supermarket of competing ideologies, a religion can claim the sanction of scientific approval, or—even better—of scientific proof, there are those who will assume it must therefore be valid." (p. 198)

There are, however, two connected questions that should be asked of this account. First, what is the basis of the claim that science is the sole or dominant epistemological authority in modernity? Second, what is the justification for linking the so-called resurgence of creationism with a more general growth of antiscience? These questions are connected, because, as I try to show, the assumption made in the case of the first—that science is the dominant epistemological authority—an assumption which, by Barker's own lights might be questioned—is itself behind the claim that creationism is resurgent.

Thus, to take the first point, it is significant to note here that Barker (1985) pointed out that,

> There has been no comprehensive sociology of this debate [between science and religion] in the present century. Of course, there is a great deal of literature informing us of the views of academics; but there is very little written about how these views are received amongst lay people. The ideas of intellectuals are very easily resisted or, even more devastatingly, ignored when they do not have some kind of resonance with significant sections of public opinion. (p. 194)

This echoes the opinion of the Royal Society about our lack of knowledge of the public understanding of science (see chap. 1). Given this, however, it must surely cast Barker's claim, regarding the epistemological authority of science, into doubt. If we know little about how the views of academics are received among nonscientists, then it seems premature to make judgements even about the epistemological authority of science. Indeed, Barker's claim that the laity are able to resist the ideas of intellectuals hints precisely at the possible presence in the wider culture of alternative bases or forms of authority to science, which in turn would cast doubt on the claim that science is the sole or dominant authority. Of course, it may be the dominant authority for some people—such as not a few academics and intellectuals, for example—but, it by no means follows that it is so for everybody, nor to equal extents, nor in all contexts.

It is, however, characteristic of the discourse of rationalization that science is presented as singly dominant and it is on this rhetoric that Barker is drawing. This becomes more plain when the second question above is considered, that of the supposed resurgence of creationism. This view is so common and so central to sociological accounts of creationism that it deserves some closer inspection.

BORN AGAIN?

Much of the encouragement to see contemporary creationism as having a recent provenance reflects the concentration of attention on those American groups that have formed in the recent past (CRS in 1963; CSRC in 1970;

and ICR in 1972, are Toumey's [1994] datings) and certainly, I would not wish to question this growth in organizational presence. Nonetheless, it is worthwhile to engage in a little reflection on the terminology employed to describe these contemporary developments.

There can be little doubt of the commonplace presence of the terminology in question. Thus, Toumey (1994) had one entire chapter entitled "The Renaissance of Creationism" (and a subheading, "The Rebirth of Creationism, 1961–1972," p. 31); Barker (1979) referred to "a growth in Creation science" (p. 181) and, referring to the post-Sputnik era, talked of the "emergence" of scientific creationism (Barker, 1985, p.184); and Gilkey (1987) commented that "in our period literalistic fundamentalism [is] on the rise." (p. 177) Similarly, Numbers (1987), in his extensive review of the history of creationism taking in far more than just the Morris-associated movements, referred to recent times as "the creationist revival." (p. 153) Nelkin (1982) was somewhat more circumspect, however, cautiously referring only to the "issues that have converged to force public recognition of complaints long ignored as merely the rumblings of marginal groups and religious fundamentalists and right-wing conservatives." (p. 19) Even so, this paints a specific kind of picture in which, in some sense or other, creationism was very much off the agenda until the more recent period.

What do these kinds of descriptions do exactly? What effect do they have? As I have already suggested, they have the effect of legitimizing the search for understanding and explanation of creationism in relation to temporally proximate factors and causes. Similarly, terms such as revival, renaissance, growth, and emergence present an image of organic development or resuscitation, with the implication that the thing in question was either in a state of atrophy, dying, or that it was merely a small seed that had yet really to flourish. It is doubtful, however, that these images of the pre-1950s situation of creationism are altogether apt. It is not difficult to present an alternative characterization of creationism in the past from which to construct a history that emphasizes continuing strength and significance, rather than the need for revivification. If this is done, it encourages a rather different view of the significance of creationism to be taken.

In effect, the issue being addressed here pertains to audiences: For whom does contemporary creationism appear as a revival or an emergence? What idea of the society in question is contained in these terms? To present creationism as the rumbling of marginal groups is to draw on a certain perspective in the representation of what, where, and who is considered central to modernity. It is only against this background that the idea of a revival makes sense.

This becomes more plain if certain other points are considered. There are three basic points, which together can be used to construct a line of continuity between contemporary creationism and that of the past. They concern the situation in the 19th-century, the situation in the United States between the Scopes trial in 1925 and the publication of Whitcomb & Morris' *The Genesis Flood* in 1961, and the situation in Britain since the 1930s. Each point will be considered in turn.

The Situation in the 19th-Century

The point to which modern debate between creation and evolution is generally traced is the period immediately following the publication of Darwin's *Origin of Species* in 1859. Even so, the significance of this event is decidedly not fully agreed.

On one side, it is commonly pointed out that evolutionary ideas were in circulation in the early 19th century, well before Darwin published and, indeed, were influential in his decision to publish when he did. Nelkin (1982) stated that Darwin's book did not so much revolutionize science as "organize and synthesize a set of ideas that had pervaded the scientific literature for more than fifty years. (p. 25)" There was, on this account, already a prevailing climate of opinion leading against traditional biblical creation.

There are other considerations, however, which seem to paint a rather contrasting picture. For one thing, it has been suggested that public acceptance of Darwin's evolutionary hypothesis has never been gained without difficulty nor resistance. Indeed, Nelkin (1982) noted that it was often members of the scientific community who were among the more vociferous objectors to the theory, as many saw their task to be the illumination of the miracle of God's work (cf. Merton, 1970). They objected particularly to what they saw as the reduction of creation to undirected chance processes, still a common argument among creationists. Within the scientific community—and biologists in particular—there was little unqualified acceptance by the late 1860s, a situation that, in a sense, continues to the present day, given the modifications to Darwinism that have been proposed in the intervening 140 years (e.g., Ruse, 1989).

A further observation made is the apparent similarity between some of the 19th century interpretations of and objections to Darwin and some contemporary critiques. According to Hodge (1988), early critics focused on such issues as the speculative, noninductive nature of the theory, the question of how complex organs like the eye could have evolved, and how a self-conscious moral being could have developed, as well as the supposed moral implications of evolution. All of these can still be found in contemporary creationist texts being referred to as problems of evolution. This might be taken to suggest some continuity of at least a core of creationist thought and support.

A further level of complexity can also be added to the issue, by the introduction of national distinctions. For example, Numbers (1987) asserted unequivocally that: "The majority of late-nineteenth-century Americans remained true to a traditional reading of Genesis..." (p. 134). He also points, however, to the growth of various attempts at accommodation between evolution and Christian teachings (Numbers, 1987), something which, as discussed previously, is also characteristic of the present day (Barker, 1980; Toumey, 1991).

Discussing the spread of evolutionary thinking in England, on the other hand, Hodge (1988) argued that, even if it appears Darwinism did gain a strong public foothold, we should not assume it was for reasons that would seem right to us today and particularly not for the truth of his account or strength of his arguments. Rather, we should look to the reasons that groups or individuals of the day may have had and, in their eyes: "the reasonableness or otherwise of ... being convinced or unconvinced at a certain time by a particular presentation of it." (p. 6) For example, Hodge (1984) suggested the possibility that an apprentice printer of the day may have accepted Darwinism—or, rather, a version of it—as it provided an ideological weapon against the "clerical hegemony" (p. 4).

In proposing such an example, Hodge—whether he might intend it or not—provided support for the general thesis being advanced here, that science in general, and Darwinian evolutionary theory as one part of science, is utilized as a resource to serve particular ends and interests in particular social contexts and conditions. It can be suggested that this may also be the case for Christianity and creationism in particular. And, if so, some continuity might be expected, as particular arguments advanced in the past may still be believed to carry weight and significance in the present, albeit not without appropriate modification.

From this, it can be seen that the attempt to look back a hundred years or so into the past should not unequivocally lead us to think that, in regard to the public understanding of evolution and its relation to Christian creationism, the situation then was altogether different from the situation now. There are grounds for emphasizing continuities between then and now, however much we might still want to highlight certain differences. Most notably, there appears to have been nothing of the legal pursuit of creationist interests that has marked the more recent period. However, such legal matters did arise in the early 20th century.

The Situation in Mid-20th Century United States

The most famous event of this time involving creationism was the trial of John Thomas Scopes in Tennessee in 1925, for having broken the State law against teaching evolution in the public schools (De Camp, 1968; Nelkin, 1982; Numbers, 1987). Although he was duly found guilty, this is widely presented as a moral (Shermer, 1991) and intellectual victory for evolution, or more generally, for science over religion and truth over dogma. Accorded particular significance in this is the role of the defense attorney, Clarence Darrow (Nelkin, 1982) and the general failure of the creationists to put together a coherent, scientific defense of their views (Numbers, 1987). Numbers commented further that the "press had not treated them [the creationists] kindly," (pp. 144–145) an image reinforced 30 years later by the film dramatization, *Inherit the Wind*. To cap the creationists' problems, their leading antievolution crusader and the main prosecutor of Scopes, William Jennings Bryan, having been roundly beaten in the courtroom debates with Darrow, promptly collapsed and died.

However, this image of creationist defeat is not easily reconciled with an apparently simple fact: The law in Tennessee was not repealed until 1967 (Shermer, 1991). Added to this, it is also widely reported in the literature that, as Numbers (1987) stated, "Darwinism virtually disappeared from high school texts ...." (p. 146) Similarly, Shermer cited the research of Grabiner and Miller, who compared such textbooks before and after the trial and concluded that "the evolutionists ... lost on their original battleground" (Grabiner & Miller, quoted in Shermer, 1991: 520).

Once again, therefore, it is possible to emphasize the continuity of creationist support and belief, especially over the period bridging the Second World War. A further point to be made here concerns the role of Henry Morris himself. Morris is the single most commonly identified figure as being at the spearhead of the 1960s creationist revival. In particular, it is often the publication of Whitcomb & Morris', *The Genesis Flood* in 1961 that is seen as breathing new life into the movement. However, a case for continuity can again be made, in that the authors seem to have drawn heavily on the work of an earlier would-be flood geologist, George McReady Price. Numbers (1987), quoting an anonymous reader, said of it, "[i]n many respects their book appeared to be simply 'a reissue'" of Price's ideas, published in the 1920s, and that it "lack[ed] conceptual novelty." (p. 150) Further, Morris had been involved in creationist groups, notably the ASA, since the 1940s and had published an earlier book also drawing on Price (Numbers, 1987; see also Livingstone, 1987).

To add to this sense of continuity, one can look at creationism outside the United States and, in particular, in Britain.

The Situation in Britain

It seems apparent from the literature that Britain has seen nothing like the same intensity of antievolution as experienced in the United States, not at any rate, since the 1860s (Marsden, 1977). Most notably, there have been none of the legislative or educational interventions sought with such apparent vociferance and success.

Here, though, there is another anomaly to disrupt any effort to see the post-1960s period as a time of revival and growth. Britain can lay claim to what must be one of the oldest continuously surviving creationist groups anywhere. The Creation Science Movement (CSM) was founded as the Evolutionary Protest Movement (EPM) in 1932, some 9 years before the ASA (Numbers, 1987). Numbers (1987: 149) described the appearance of this "small but vocal group" in the 1930s as sudden and as catching "nearly everyone by surprise." This statement is notable for its use of an extreme case formulation (Pomerantz, 1986) involving what Potter, Wetherell, & Chitty (1991) called quantification rhetoric. The "nearly everyone"—especially presented against the prior description of EPM as small— invokes a sense of abnormality, of a situation in which acceptance of evolution had become so much taken for granted by the vast majority that any questioning of it would be entirely unexpected. Once again, therefore, the representation at work is one that uncritically assumes unquestioned and unproblematic acceptance of evolution as the norm. Against such a background, the establishment of EPM inevitably appears sudden and surprising. Yet, it is the representation that makes it appear so.

Looking at the founding figures of EPM, however, does not put one in mind of marginal radicals, so much as mainstays of the British establishment. According to their own sources (Geering & Turner, 1989; Turner, 1982), EPM was founded by Captain Bernard Acworth, with the most active assistance of Douglas Dewar. Other members of the founding Advisory Committee included Dr. Basil Atkinson, then Under Librarian at Cambridge, and Dr. James Knight, Vice-President of the Royal Society and "a scientist of distinction" (Turner, 1982). Dewar was an amateur ornithologist who had presented papers to various learned bodies and had published widely on natural history. He seems to have been led to his rejection of evolution more on the basis of his observations than by any predisposing religious beliefs (Numbers, 1987). Other founding members, however, were more noted for their evangelism. They were "representative of several denominations" (Geering & Turner, 1989) and the Movement continues to reflect these characteristics, being both scientific and evangelical, and somewhat ecumenical.

The stated aim of CSM was and is to combat evolutionism on "both scientific and moral grounds" (CSM editorial). Once again, therefore, it is possible to bring out the apparent continuity between earlier and contemporary forms of argument. Emphasis continues on the unscientific nature of evolutionism, its lack of empirical support (for example, the absence of transitional forms in the fossil record), its inability to explain significant empirical phenomena (such as apparent design in nature and the existence of the human moral sense), and its immoral influence on society at large.

As noted, EPM/CSM has had relatively little success compared to its counterparts in the United States, certainly on an institutional level. Equally, there is no reason to think that its views are widely known in Britain,[5] despite the energetic efforts of members to spread their views. It is a small group, with a membership standing at some 1,300 in November, 1989 (D. Rosevear, personal communication). Of these, approximately half are graduates, mainly in science, with around 200 higher degrees. There appears to have been some effort to expand their base of support over recent years, with "many more meetings ... being held and tapes, slides and videos ... added to our book lending facility" (Geering & Turner, 1989, p. 4). Similarly, a separate, but closely linked organization, the Creation Resources Trust, was established in the early 1990s, for the sale of books, slides, videos, and tapes, and the publication of "attractive creationist papers for children and teens" (CSM publicity notice). These pamphlets are distributed by CSM in addition to its regular bimonthly journal, *Creation*, and the

---

[5]An impressionistic indicator of this: I have been discussing and teaching the subject for about 9 years now and have rarely encountered anyone—colleague, student, or acquaintance—who had heard of creationism or creation science. A typical response on being informed about them is a combination of incredulity and amusement. There is some possibility that this may change, however, with the continuing growth of evangelical Christianity (Bruce, 1984; Davie, 1994; Kepel, 1994) and creationism appearing in the British media (e.g., *Everyman: Science Friction*, BBC1, 8 September, 1996).

pamphlets used in the research for this book. These materials seem to have been the major means of circulating their ideas from the outset.

The change of name, from EPM to CSM, occurred in 1980, indicating, according to Turner (1982), a shift from the "largely negative" criticism of evolutionism in earlier days to a "more positive position demonstrating the complex nature of the natural world with its wonders and the impossibility of it evolving." (p. 15)

The point of all this, then, is not to deny that there may indeed have been some significant developments in the organization and, perhaps, the attempt to systematize, or at least disseminate, the ideas of scientific creationists since the 1960s. Nor that there has not been some new resort to legislation, involving significant changes in strategy, suggesting a greater accommodation with the secularized political sphere (Shermer, 1991). Nor is the point just to show that it is possible, without too much difficulty, to make a case for the continuity of creationist belief and support. If this were the intent, it would simply be asserting the priority of one partial perspective over another.

The point is, rather, to highlight that the description of recent developments as a revival, or a renaissance, or an emergence adopts a rhetoric that imports a certain set of assumptions and imposes a certain model of societal order. To see this better, all that is required is to imagine a simple turnaround and view the events of the period since the early 1960s not as a revival of creationism, but as a revival of the crusade for evolution. It is not creationism that has experienced rebirth, it could be said, but the demand to replace it with a monopoly of evolutionist thought and teaching. To do this, however, involves questioning the assumption that modern society is first and foremost a scientific society, meant in the sense of being a materialist and secular society. It also involves conceding something to the creationists: their *right* to criticize science. To see the issue as not one of a rebirth of creationism, but a new marshalling of evolutionist forces, requires adopting this different kind of stance. It requires seeing things from the creationist point of view, in which what to them might have been seen as their stable and congenial society, built in part around religious convictions, is suddenly under assault from this external presence, which in their view has nothing to do with them.

Now, as previously mentioned, this is not a view that is altogether absent from sociological accounts. It appears in the form of seeing creationism as a negative reaction to the imperialist expansion of Big Science and the Big Brother military–industrial complex that goes with it. My point, however, is that this perception of things sits unhappily beside the rhetoric of rationalization in which science is presented as already dominant.

More generally, this tension can be seen as a consequence of the over-thrusting, as it were, of the dilemmatic of science. According to the one view, science is a universally valid form of knowledge that imposes an irresistible truth, which to resist is both ignorant and futile; according to the other, resistance is to be expected, because science is no more than one type of socially bound form of knowledge, which has no inherent right to dominance or singular faith. The accounts of creation science considered above, play out this tension, trying to satisfy both sides of the dilemma, in accordance with a sociologic that, on the one hand draws on rationalization, but on the other, finds itself confronted by the anomaly of creationism, which it tries to explain by resorting to the principles of the sociology of knowledge contained in the second view.

This whole complex appears in a particularly marked form in Gilkey's (1987) account, which I use by way of a summary of the foregoing critique. Although Gilkey is a theologian by training rather than a sociologist, I include his account of contemporary creationism here as it is strongly informed by a sociological orientation. This seems to justify treating it for its sociological features and claims, rather than its theological ones.

GILKEY AND CONTEMPORARY CREATIONISM

Gilkey's (1987) argument is almost classically Weberian, though with a key modification more reminiscent of Habermas (see chapter 6). He sees creationism as a perfectly understandable response by a religious commu-nity to a perceived attack on their traditional beliefs by the advance of a bureaucratically embedded science that pays little attention to local con-cerns in its inexorable advance. The advance of science is a direct conse-quence of its greater explanatory power, on a material and naturalistic level. However, science is, in Gilkey's view, incapable of addressing the moral–practical concerns of ultimate meaning and purpose to which tradi-tional beliefs are directed. As science has advanced, therefore, it has displaced these beliefs, but left nothing in its wake that can adequately substitute for them.

  The view that religion as such was an aspect merely of past, prescientific
  societies, and thus of ignorant and vulnerable societies, has been a part of
  this scientific and technological self-understanding characteristic of modern

cultures, an aspect in effect of the founding "mythos" constitutive of modern scientific culture. This view understood the major cultural activities of the past as if they had been solely cognitive endeavours, and thus it saw all forms of past cognition as either prescientific or pseudoscientific, based on the pattern of its dominant form of knowing. Thus the religious myths of the past were seen as prescientific efforts to explain natural events, to provide the sort of important but limited understanding science provides. For this reason, it was reasonable to believe that all forms of religious "knowing" would dissolve away as science itself was developed and replaced them. (p. 176)

In presenting this image of rationalization as the mythos of modernity, Gilkey clearly distanced himself from this as an explanatory model. He presented the view as mistaken, because religion—or, rather, the mode of thought of which the religious is one expression—can never really disappear from human existence. It is as necessary to us to consider the ultimate questions—those that address the "transnatural (or personal) factors" (p. 173) behind the material realm of science—as it is to deal with the material world itself. Thus, creationism—and the rise of "literalistic fundamentalism" (p. 177) more generally—is no more than ought to be expected. Indeed, they are partly in response to the very development of science itself. The previous quotation continues:

Unexpectedly ... the historical development of this scientific culture has itself shown the one-sidedness if not the falsity of this latest cycle of scientific myths establishing that culture. Out of that culture's own advances have arisen dilemmas which raise religious questions and which call for religious answers. (p. 176)

Dilemmas, that is, addressing the nature of our collective existence, which, even as our material power has waxed, seems itself just so much to have waned. Into this widening breach have stepped the new fundamentalisms, in an effort to appease the anxiety of a society deeply at odds with itself.

Equally, it is no surprise if some of these fundamentalisms seek to wrest the mantel of scientific authority from the material side. The reason for this, noted Gilkey (1987), is because, "science represents for our culture the paradigmatic and so sacral form of knowing" (p. 170).

Herein is the rub. Gilkey described science here in terms akin to those encountered in Gellner (chapter 1). Science underwrites so many of the central institutions of our society—"medicine, technology, social policy, defense" (Gilkey, 1987, p. 170)—that its presence is inescapable and, perhaps, irresistible. "Each form of sacred knowledge has for very understandable reasons a sacral aura, symbolized in our case by the white coat and by the super prizes bestowed on scientists." (p. 170) It is no surprise, then, if this almost religious quality should attract bids from other, more traditional, religions to a share of its status.

But has Gilkey not already told us that it is a myth of modern self-understanding that science can displace religion? If so, how can it be that science can be considered sacred in this way? This is the *myth*, not the *reality*.

The confusion in Gilkey's account here becomes more manifest in his further discussion of the relation between science and the wider culture in modernity. The above view presents us with an image of modernity as thoroughly steeped in and dominated by science; science is sacred (cf. Whitley, 1985), and it is fundamentally different to other forms of knowledge or belief. It has, says Gilkey (1987), "logical limits ... as a mode of knowing"—the cognitive mode, concerned solely with the objective and sensible (p. 173). Thus, its dominance has left a moral vacuum that is now being filled by various revisionist religions, which take a fundamentalist form to compensate for the materialist fundamentalism characteristic of science's self-understanding.

Yet, Gilkey (1987) also had this to say about modern culture: "A ... consequence of an advanced scientific culture is that science permeates down to and shapes all levels of modern society. *In turn it is, therefore, taken over and shaped by all levels.*" (p. 167; author's emphasis)

Gilkey drew a parallel with the history of popular religion, which arises when a belief has also experienced such permeation:

An established religion then takes on, *as a part of itself*, local, age-old, often deviant or bizarre forms (syncretistic forms) as a result of its mingling with the whole range ... of the culture .... *Thus what we here refer to are forms of modern science*, however they may horrify the American Association of the Advancement of Science .... (p. 168; author's emphasis)

... or, I might add, the authors of the Royal Society report into the public understanding of science! The forms Gilkey considered are such things as the strongly nationalist sciences to be found in Nazi Germany, Stalinist Russia, Shinto Japan, and elsewhere, but also the more local sciences, such as creation science.

Now, my point is that this second view of creation science is really very different from the first. The first view presents creation science as essentially a bastard parasite. It is an outcome of a situation in which traditional beliefs have lost their credibility with the rise of science; they no longer cut the cognitive mustard. However, they are still the only things people can resort to for the moral guidance they need. So they hang around, somewhat limply, until some bright spark gets the idea of boosting them with a shot of artificial scientific testosterone.

The second view almost completely reverses this image. Here, science is no longer presented as dominating the scene, but, rather it appears as only one among a number of forms of knowledge, which have at least equal status with it, or even, as in the case of some nationalisms perhaps, hold sway over it. In fact, I want to suggest, this second situation is even more radically adrift from the first than this implies: For, if what Gilkey says about science taking on these other beliefs "as a part of itself" is true, then the whole construct of a division between science and other forms of knowledge collapses—at least in the terms that Gilkey has built. The first view—creationism as parasite—is only conceivable if a firm distinction between science and religion is made. It is, indeed, an outcome of making such a distinction, because the distinction itself predicates that science is beyond the kinds of social processes that religion is said to both express and support.

Thus, although Gilkey sought to distance himself from the mythos of modernity as a purely technical kind of civilization—that is, he recognized this for the rhetoric it is—he did so incompletely. He continued to draw on this understanding himself in his attempt to construct a distinction between science and other forms of knowledge. At this point, therefore, he became persuaded by the rhetoric.

In advancing the alternative view of creationism as an outcome of syncretistic processes, however, Gilkey moved closer to seeing science as a cultural resource. From this position, there is no need to attempt to make distinctions between science and other types of knowledge or belief; rather, all are seen as resources that are potentially available for the purposes of constructing explanations and advancing and deflecting arguments. There is no difficulty in seeing science as simply one such resource (or, better, as a multiplicity of such resources) that may or may not be found in particular social locations used alongside resources ostensibly drawn from other ideological repositories. And one should anticipate, as a complete part of this process of ongoing public argumentation, discussion, and disagreement about the nature of science and its relation to other forms of knowledge or belief—which might well include making use of distinctions between science and religion of just the sort that Gilkey himself presented. The use of such distinctions, then, can be treated as a topic for study, not a resource for the analyst to employ.

To be clear: My argument is that there is a confusion in Gilkey's account arising from his contention that scientism is a myth of modernity, only then to go on to argue as though this myth were true. In other words, Gilkey himself stated that the view that science can displace religion is a myth, but he also claimed that science is considered sacred in modernity. My point is that the view that science is sacred is precisely the myth that Gilkey attempted to undermine.

Moreover, in developing his counterargument, he presented an account of science in modernity, which itself undermined the claim that science is sacred. If science were sacred, then it could not be "taken over and shaped by all levels" within the wider culture. For this to happen, science must be viewed as only one among a range of types of belief (i.e., cultural resources).

Similarly, in stressing that there are logical limits to science, Gilkey adopted the discourse of rationalization, which presents science in monolithic and reified terms. Yet, this conflicts with the representation of being open to shaping by ostensibly nonscientific discourses. On the one hand, then, Gilkey said that science has clear boundaries, demarcating it from nonscientific things (such as morality and nationalism); but on the other, Gilkey maintained that it is "taken over and shaped" by these other things, in which case its boundaries become unclear and can be expected to be (as he himself suggested) historically and culturally variable.

I would not expect Gilkey himself to be fully happy with my contention that his position is, at least in part, consonant with the perspective of cultural resources. As a theologian, Gilkey might well wish to demarcate the religious from the nonreligious and strive to protect it from contamination in the same way scientists seek to protect science. From my perspective, however, religion should no more be considered insulated from cultural (re-)shaping than is science. As much as he presented a reified conception of science, Gilkey would be likely to present a reified view of religion, having clearly defined boundaries demarcating it from other things (such as science). However, in presenting a view of science as open to shaping by nonscience, he has in effect undermined the very reified categorical expression he adopted. Naturally, I would want to hold the same basic point to apply to any categorization of religion—but it is not crucial to my argument against Gilkey that this be held at this point. My argument is that he finds himself presenting an understanding of science that accords with the view of a cultural resource despite himself. This is because the reified categories of rationalization are empirically inadequate, but also theoretically internally unstable, because of the central dilemma of science in modernity.

It might also be perceived that my argument is more applicable to mainstream sociologists, such as Nelkin and Barker, rather than a theologian such as Gilkey. However, whether theologian or not, Gilkey argued from a sociological standpoint. In so doing, I contend, he adopted the language of mainstream sociologists, at least for some of the time. Perhaps because he is a theologian, his thinking is more open to an alternative discourse as well, at least so far as science is concerned; but this alternative discourse of cultural resources, works to undermine the discourse of sociological rationalization. Thus, Gilkey is, in my view, a better illustration of my argument than Nelkin, precisely because in his account, the tensions between these two discourses achieve a much fuller expression.

The previous analysis, then, leads to certain questions that might be asked of creationism. What understandings of science do creationists employ? How do these understandings stand in relation to those used by academic commentators on creationism? Are they the same or different? And what exactly are they used for? These questions are addressed in chapter 3.

CONSTRUCTING CREATION SCIENCE

What I have tried to show in this chapter is that creation science is a category that appears in sociological texts as a construct designed to fit into a certain would-be explanatory model. In this model, science appears as a monolith, an entity that imposes a certain logic on the social context around it, whether it be in the behavior of members of the scientific community, or in the broader self-understanding of modernity. Further, this thing science is contrasted with other kinds of beliefs and the actions said to be associated with them—especially beliefs and actions of a moral–practical kind that are, in turn, said to be characteristic of traditional, or prescientific beliefs. The displacement of these beliefs by this science has generated a certain kind of rupture in the social sphere which in turn has provoked the rise of creation science in an attempt to, if not heal the rupture, at least fill the gap. What is crucial to this is that science appears monolithically and that creation science appears as a recent response to it. However, not only is it possible to construct an account of creationism that contrasts with this view of it as recent, but also it is notable that an alternative version of science—as a socially contextual multiplicity—keeps intruding into these would-be explanations.

This shows that creation science is a construct; it is constructed to fit a view of modernity as a scientific society, one in which so-called traditional beliefs no longer have coinage or credibility. That it is possible to construct an alternative version of creationism as a continuous tradition dating back to the 19th century suggests that this view of modernity is doing more than merely describing. Rather, in describing it is arguing, in order to make its argument it constructs an absolute contrast between creationism and science. However, the stability of this contrast is continuously being threatened by the alternative view of science that keeps intruding. In this view, science is no monolith; rather, it is socially relative and multiple. What this indicates is that the explanatory model is inadequate; it cannot account for creation science without collapsing and without drawing on the resources of the alternative view, of science as a cultural resource, in a vain attempt to shore up the gap.

Creation science is an anomaly that stretches the resources of the discourse of rationalization to the point where they fail. In failing, use is made of the resources provided by the view of science as a cultural resource instead. This is because these two sets of resources are each contained in the core of the sociological view of science, which itself expresses the dilemmatic nature of science in modern culture. This dilemma poses a contrast between science as a universally valid form of knowledge, effectively outside of the social, and science as shaped by social determinations. In the next chapter, I show that this same dilemma is to be found at work within creationist texts, where it is used to provide a range of resources with which to undermine evolution.

3

Science—After Its Kind(s)

Chapter 2 considered the construction of *creation science* as an object in sociological accounts, specifically as a recent phenomenon. This chapter considers its construction as a singular, unified entity. As argued earlier, the rationalization hypothesis advances a view of science as a monolithic entity that displaces other forms of thought and belief, also presented monolithically. Thus, science stands contrasted to creation science as one thing to another. Such a view implicitly raises the matter of boundaries. Somewhere, a clear line must exist that demarcates the thing, science from the other thing, creationism. Thus, it is characteristically found in sociological accounts of creationism that some such boundary line is drawn—and drawn *on*, in the sense that such boundaries often seem themselves to be derived from resources provided by certain philosophical or sociological views of science, and once drawn, any such boundary is then used as a point of reference in the kind of account of creationism that is advanced (Gieryn, 1983; Gieryn et al., 1985; Lessl, 1988; Prelli, 1989a; Taylor, 1996).

However, it is also notable that the boundary lines identified vary; that is, different sociologists employ different grounds to demarcate science from creation science. Equally of importance, they point to differences within creationism itself with respect to the kinds of understanding of science that creationists themselves advance. This shows not only that creation science is a construction of the accounts given of it, but that these constructions themselves draw upon a range of models of science—models, crucially, which are also used by creationists in their attempts to undermine the theory of evolution. In other words, sociologists use these models to undermine creationism, but creationists also use them to undermine evolution!

Looking at this feature of creationists' discourse also provides a useful way to introduce the basic techniques of discourse analysis (DA). I will show step-by-step how such an analysis can be accomplished by applying these basic techniques to the use made of models of science within the pamphlets published by CSM. My own model of DA is drawn from the outline provided by Potter & Wetherell (1994).

MODELS OF SCIENCE AND THE
DILEMMATIC OF SCIENCE

First, I need to flesh out what I mean when I talk of models of science, an outline follows of the significance that I attach to their usage. The general argument here, as outlined in chapter 1, is that there is a shared discourse of science within modernity, a discourse marked by internal tensions, which arise from the dilemma posed in the modern attempt to understand science as both a universally valid form of knowledge—and therefore one in some way above and beyond social limitations—yet, also a type of knowledge uniquely connected to modern society. At its simplest, this dilemma can be represented as a debate between two fundamental views of science, the asocial universalist and the socially localized. This contrast in views can be used to locate a range of more specific models, or representations, of science, which purport to describe its essential features and mode of operation.

This in itself is so precisely because of the contrasting nature of the views in question. The views embody a basic rhetorical contrast between the general and the particular. As Billig (1985) argued, this contrast is never altogether resolvable as:

> a statement asserting a categorization can be opposed by a statement asserting a particularization. If there is always a way of distinguishing the particular instance from the general category (because it possesses particular properties), then it is always possible to counter the categorization by the special case (and vice versa the reverse is also true). (p. 97)

This gives rise to arguments attempting to defend the one view against the other. In the case of science, then, the tension works through to the level of its inevitable grounding in human action. General knowledge claims are made always from particular occasions of scientific work (Potter, 1996; Taylor, 1996). It is always possible, then, to dispute the generality of a claim by appeal to its particularities and especially those of its human and social contextual features. As will be seen in the next chapter, there are established means available within the discourse of science for managing this tension. It is also notable that, as this tension works out in particular cases of scientific controversy, a type of "experimenter's regress" (Collins & Pinch, 1993) occurs, in which the argument tends to shift away from the content of knowledge claims to the level of methodology, focusing instead on experimental procedures and practitioner competence. In the case of controversies on the borders of science, although charges directed at this methodological level may still be made, debate shifts further still, to the metalevel of the nature of science in general. The precise reasons for this shift are unclear, although Gieryn (1983; Gieryn et al., 1985) made a strong case for seeing them as part of the concern to build and maintain a level of occupational control over the "market for knowledge" (Gieryn et al., 1985; cf. Lessl, 1988).

The intention here, however, is not to seek explanations at this level. Remember that, although the elaboration of models of science has primarily been the work of the class of intellectuals involved with knowledge production whose interests might be served by the construction and defense of boundaries, the fundamental tension arises also within the wider commonsense lifeworld of modernity, albeit discursively elaborated to as yet undetermined extents and in unidentified forms. Boundary demarcation is indeed a practical problem for scientists, as Gieryn has shown. The claim here, however, is that it can be a practical concern for nonscientists also, whatever the specific reasons for this may be. It is important to recognize in this respect that the authority of science in modernity should not be assumed but treated as an accomplishment and, as such, as something that may be expected to vary contextually, with degrees of success. This can be seen in the case of creationism. Thus, the basic contrast, between universal and local representations of science, is present in their discourse and is more or less elaborated in the form of the adoption of a range of different models of science. These in turn share a basic identity with the range of models of science drawn on by their sociological critics—although, interestingly enough, in some respects, creationists display a wider range of representations of science than these more orthodox commentators.

The two views of science work through, then, into the elaboration of a range of different models of science—or, at least, I am treating the expression of particular representations of science as indicators of one or other of these basic views[1].

The view of science as asocial is one that, broadly speaking, is in keeping with the idea that there is a world of facts and/or fundamental laws, which is essentially outside of human influence and that science aspires to describe

---

[1]A more conventional treatment of different models of science can be found in Chalmers (1982).

and represent. Because these facts or laws are deemed to be everywhere the same for all peoples in all times and places, then, to the extent that science is successful in its efforts to describe them, so it can claim to be universally valid and objectively true. Correspondingly, this view includes representations of science as a body of knowledge, or as a process of induction (i.e., reasoning from the particular—say, a collection of observations or other facts—to the general, such as a theory, explanation, or law). It also includes hypothetical–deductive models (i.e., reasoning from general theories to particular, empirically testable hypotheses), for, however sophisticated the falsification, it still ultimately assigns the determination of validity to events occurring in a world existing independently of human beliefs. Also included here are representations that point to criteria of internal logical coherence of theories in the demarcation or assessment of science, as these also appeal to principles of universal validity above and beyond specific human subjects or social locations. In these cases, logic, reason, or rationality are treated as just as constraining of human thought and action as are facts.

In contrast, the socially localized view of the nature of science links science to the human and social context of its appearance. It says that science is a matter of local conventions and, as such, that it is in some way dependent on one or other feature of the community of people who are involved in doing the science. From this point of view, science is not simply a matter of the production of facts, nor just about making sense or being reasonable; rather these processes themselves are seen as secondary in importance to personal and collective conditions that work to constrain the way in which facts are produced and theories are assessed. This view is represented in such broad notions as that science is open to subjective influences, or is a matter of interpretation or bias. It is also found in more subtle forms, such as Merton's (1968a) normative system model (see chapter 1), and in Kuhnian-inspired notions of paradigms (Kuhn, 1970), with the attendant idea that the normative order of the scientific community extends beyond the testing of knowledge-claims to the actual means of testing and cognitive frameworks of understanding themselves. A step further still brings in a more thorough sociology of knowledge model of science, especially the strong program in SSK (Bloor, 1976), in which all aspects of science are held to be relative to social, historical, and cultural factors (see also, Hollis & Lukes, 1982; Wilson, 1970).

The general point to be taken from this is that these various models can be thought of as manifestations of a simple, basic contrast that informs understandings of science throughout modernity and provides a set of rhetorical resources—a repertoire of models of science—enabling either view to be defended or attacked with equal effectiveness and persuasiveness. At least, this is what I want to suggest can be seen at work in creationist texts and in texts on creationism. It is to these that I now turn, beginning with sociological commentaries on creationism.

CREATION SCIENCE AS NAIVE SCIENCE

Reading sociological accounts of creationism, one might be forgiven for thinking that creationists are pretty naive about science. They are commonly presented as misunderstanding it, both in general principle and in particular detail. Nelkin (1982), for example, said: "[Creation scientists] use the language of science but seem to understand little about its methods and underlying assumptions. Indeed, the creation controversy illustrates two common beliefs about science that bear on its acceptance: (1) that science can be defined as a collection of facts, and (2) that it can be evaluated in terms of its influence and implications" (p. 188). Similarly, they "view science as an inductive and descriptive process and poorly comprehend the function of theories and models as useful instruments for prediction." (Nelkin, 1982, p. 76)

Based on these descriptions, then, it would seem that creationists unequivocally hold a universalist view of science, seeing it as a simple description of the facts of reality, with theories inductively derived. At the same time, however, they are also guilty of introducing other criteria into the assessment of science, thereby contravening the logical limits of its boundary.

Similar observations are made by others. Toumey (1994) presented creationists' views of science against a background picture of modernity as dominated by what he called "the trivial model" of science. He defined this as one empty of "substantive content" of the knowledge base of science, but which relies on "the artful deployment of the easily recognizable symbols of science" (Toumey, 1994, p. 142). Further, he contended that United States fundamentalists, at least in the early part of this century, "were very much in favor" (p. 21) of a 19th-century model of science, which he calls "the Protestant model," that drew in part on "Baconian empiricism" (p. 19). Thus, he also seems to link creationism with induction. In similar fashion, the philosopher of science, Kitcher (1982) claimed that creationists assess evolutionary theory using a naive falsificationist model of science; for example, they maintain that evolutionary theory has been falsified by the failure to find transitional forms in the fossil record. Others also discuss creationism in relation to such singular models of science (e.g., Cavanaugh, 1985; Dolby, 1987; Ruse, 1982).

An important theorist to consider in this respect is Taylor (1992, 1996). Taylor's general position with respect to the demarcation of science is closely allied with the present argument, seeing it as a social, and precisely a rhetorical, construction: "The meaning of science, as a set of social practices, is constructed in and through the discourses of scientists as they respond rhetorically to situations in which certain of their social, technical, professional ... interests are problematized ...." (p. 5)

Thus, science is "what it is rhetorically demarcated and authorized to be" (Taylor, 1996, p. 6). Taylor elaborated this argument through, among other things, a detailed analysis of the creationist controversy in the United States, considering both the creationist demarcation of science and the resulting response from the orthodox scientific community. His central point was that this response is itself part of the problem, serving to perpetuate creationism, because it fails properly to appreciate the kind of public appeal that creationism has. This appeal is partly to do with "a mistrust of detached technical expertise," but also results from the adoption by creationists of a "quasi-Baconian inductivist demarcation of science" (Taylor, 1996, p. 140; cf. Taylor, 1992), which resonates with the "empiricist folk epistemology" said to be characteristic of modern culture—a claim Taylor (1996) has adopted from Cavanaugh (1985).

Both Taylor's general rhetorical stance and his detailed analysis of creationist argumentation lends much support to the present account. Despite this, however, I must take issue with his claim that creationists consistently adopt a Baconian–inductivist model of science, as this does not accord with the range of models of science I have found in the discourse of CSM. Moreover, Taylor's own material does not quite fully fit the picture he paints, in particular, his analysis of the scientific response to creationism noted that both sides in the controversy have called upon Popperian criteria of falsification to undermine the scientific status of the other. Pointing out that Popper had at one time maintained that Darwinism was not a falsifiable theory, Taylor (1996) noted that "creationists seized on this seeming inconsistency, arguing that evolution was, at best, no more scientific than creationism." (p. 164)

In other words, creationists mobilized the same model of science as was appealed to by many scientists—a model of science that was falsificationist, rather than inductivist (cf. Kitcher, 1982). From the present perspective, it is important to emphasize this. It demonstrates the rhetorical flexibility displayed by creationists and shows that they are not simply beholden to a single model of science. It also, incidentally, calls into doubt the claim that their inductivism is grounded in something called an empiricist folk epistemology. There is already evidence to suggest that this is an oversimplified view of folk epistemology compared with studies in the public understanding of science discussed in chapter 1. Moreover, it appears to be an unwarranted denial of mundane reasoners' rhetorical skills. Oddly, this is something that Taylor himself, at times, appears to concur with, when he points out that members of the public "do not react simply to technical content, but to a complex of contextual, institutional, and personal representations of science" (Taylor, 1996, p. 138). However, whilst he appears to recognize that this view exposes the artificiality (i.e., the constructedness) of the distinction between the technical and the social, yet he does not seem to hold to this firmly enough in the case of creationism. Thus, one of the things that makes creationism so troublesome for any monolithic construction of demarcation criteria is precisely that they deconstruct the distinction between the technical and the social (or, more precisely, the moral) in systematic, albeit variable, ways.

This, in fact, is commonly presented as a further form of misunderstanding, as it breaches the supposed logical limits of science. Gilkey (1987), for example, saw creationists as overstepping the boundaries of science, mixing the moral with the cognitive. He further saw the "logical limits of science" as confined to "the domain of finite, material, and objective causes" and to "objective, invariable, and sensible" processes (p. 173). Thus, the fact that creationists import religious and other moral factors into consideration shows them to be contravening these limits. In mobilizing this contrast, Gilkey adopted the typical strategy of sociological commentators. In this strategy, creationists are said to understand science in one specific way, to which the sociologist offers a contrasting understanding and so exposes the mistakes and/or naivety of the creationists. Thus, just as Gilkey argued that creationists contravene the logical limits of science, so Nelkin (1982), having defined creationists' understanding of science in empiricist terms, advanced her own, broadly normative, model: "Scientists accept theories and teach them, not because they represent 'truth,' but because they are accepted by the scientific community as useful explanations of reality ...." (p. 186)

In similar vein, Kitcher, having said that creationists adopt a naive falsificationist model of science, constructed a critique based on his preferred sophisticated falsificationist model. Likewise, Toumey proposed a contrast between the trivial and 19th-century Protestant models of science, and "the secular model of the European Enlightenment, explicitly grounded in rationalism and naturalism" (Toumey, 1994, p. 19)[2]. Further, his stress on the symbolic usage of science seems to resonate with Nelkin's claim that creationists adopt the language of science without real understanding.

As can be seen, then, there appears to be a fairly typical argumentative strategy adopted by orthodox commentators, in which creationists are characterized as holding a particular model of science, broadly of a universalist form, which also breaches logical limits by importing moral criteria. Having then set creationism up, it is thus in a position to be knocked down by the simple mechanism of introducing a contrasting understanding of science of the appropriate kind—often said to be more sophisticated and logical.

It must be said that from a practical point of view, this has proven to be a very effective strategy for scientists in public forums. As a number of analysts have shown (Gieryn et al., 1985; Prelli, 1989a), the use of the normative model was a highly successful strategy in countering creationists' legal challenge for equal-time in the Arkansas hearing. In particular, it persuaded presiding Judge Overton of the illegitimacy of the creationist claim to scientific status (Prelli, 1989a; see also Overton, 1982).

For the purposes of enhancing sociological understanding, however, the strategy is more questionable. In particular, it is compromised by the variability displayed in the characterization of creationists' understanding of science. Thus, as we have just seen, creationists are depicted sometimes as inductionists, sometimes as naive falsificationists, and sometimes something else. To add to this, Barker (1985) noted of creationists themselves that,

> ... while it is sometimes claimed that science *proves* Genesis correct, there are those who will claim that it is because science cannot tell us with *certainty* whether Genesis or evolution provides the true explanation of man's origins that both should be given equal time in science classes. Thus, some creationists will claim an epistemological *relativism for science* at the same time [as] they denounce ethical relativism in favor of *absolute moral standards.*" (p. 199; emphases original.)

[2]It must be acknowledged, however, that Toumey's (1994) position is less unequivocal than this might suggest. Although he seemed to link creationism especially to the trivial model of science—that is, one that is content to appropriate the symbols of science, shorn of content—he also observed that contemporary creationism is "a rich, complicated, and varied system of knowledge, values, and beliefs ..." (p. 143). He also said, of the ICR and CRS in particular, that they are "most concerned about idioms of scientific respectability—for example, secular scientific credentials, quotes from Karl Popper, Kuhnian paradigms, and so on" (Toumey, 1991, p. 694). This leaves it unclear if he is suggesting that the use of such quotations and the possession of secular credentials should be perceived as merely symbolic; if so, one wonders of what a nonsymbolic use of science might consist! In any case, it is doubtful that the notion of a trivial view of science does sufficient justice to the complexity of creationism.

What is to be made of this variation? Two points concern me. First, there is an evident need to try to establish what view of science creationists do in fact have. As will shortly be seen, they seem to have multiple views of science. The question, then, is to understand why such variation is displayed in their writings. I argue it is because different models of science may be mobilized to different ends, that they serve different rhetorical purposes within the context of different arguments. The question, then, in a sense is more fundamental: Why do such different models of science exist? As should be clear by now, my answer is that it is because of the dilemmatic of science, which provides the source of these rhetorical resources.

This is supported by the second point, which is that these same models of science appear to be used also by sociologists discussing creationism. These sociologists draw on creationist texts in order to construct creationism in a particular way; having so constructed it, they then proceed to knock it down by themselves exploiting the resources provided by the dilemmatic of science. Both groups, then, are behaving in a similar way: The sociologists construct creationism in a way that facilitates their critique of it; and the creationists construct evolution in a way that facilitates their critique.

Let us now look more closely at creationists' representations of science.

THE EXPLOITS OF MODELS

What follows is an analysis of the models of science found in the pamphlets published by the Creation Science Movement (CSM; see also chapter 2). CSM has been publishing these for several decades now. The earliest dated issue I have is from 1958, and that is a revised edition. They are cheaply produced, mostly consisting of a thin card or sheet of folded A4, with printing on either three or four sides, sometimes with a cover picture drawing and/or other illustrations. A small number have more pages (up to 14 sides) and contain much more detailed illustrations. Most pamphlets provide a brief discussion of a single topic, though a large minority offer more encompassing expositions. Mainly, they discuss naturalistic issues deemed relevant to the ongoing critique of and argument with evolution. They cover the gamut of scientific disciplines, from physics to biology, and on to archaeological, anthropological, and sociological matters, though with notable concentrations on geology–paleontology and biology (including a large minority of natural histories). A small number are focused solely on biblical exegesis, although biblical references appear often elsewhere.

Although other sources were also read as part of the background to this study—including as many books by other creation scientists as could easily be obtained,[3] as well as books and articles of commentary and critique by members of the orthodox scientific (natural and social) and Christian communities—the pamphlets were found most useful as the focus of analysis. They provide sufficiently discursively elaborated statements of creationist views and arguments as to bear analytical attention, yet are sufficiently brief to enable a wide coverage of expression from different authorial voices. Also, although having the preconstructed characteristics of any written composition in essay form, they are relatively spontaneous statements, in the sense of being undirected by the analyst. This is not to say that such closer techniques as interviews, for example, might not also elicit highly valuable material; indeed, it is to be hoped that future studies might provide such data to supplement the present analysis (e.g., Coleman & Carlin, 1996). Nonetheless, the pamphlets are worthy of attention in their own right. They provide a bridge over the years of existence of EPM/CSM and, thus, display features of the continuity of creationist views, while also advancing more recent developments and additions to their corpus of antievolution arguments. Finally, they have one compelling feature: Like fast food, they are most conveniently prepackaged!

A sample of these pamphlets was obtained directly from CSM and analyzed using the techniques of discourse analysis (DA). This raises the procedural question: How exactly does one do DA? Although there are no preset rules for this (Potter, 1996; Potter & Wetherell, 1987), some indication of how to go about it can be given by example.

One immediate answer is that DA is a lot more difficult than it might appear to be from reading other researchers' studies. At its best, following an analysis can seem to be almost a disclosing of the obvious. This somewhat odd formulation (does the obvious require disclosure?) is deliberately chosen, because although DA may seem to focus on the superficially obvious features of a text, what is made of those features when they are highlighted so explicitly can prove far more revealing. DA tries to observe what is displayed in a text. This may seem either trivially obvious, or an absurd oversimplification of the complexities of the reading process—or

---

[3] Apart from those published by CSM, these are not easy to obtain. Publications by U.S. creationists are particularly difficult to obtain. They are rarely stocked in British University or public libraries and, although obtainable through Interlibrary Loans (the main system I have used), I have experienced considerable delays. Given this, I have largely sufficed with material available through CSM, together with the critical secondary literature by orthodox commentators, which is much more easily obtained (see footnote 1, chapter 2). It is interesting to note in this respect the creationist claim that their views are systematically excluded from the institutional forums of the scientific community.

perhaps both! At one level, every reading of a text, assuming it does not completely ignore it, must observe what is displayed in it—how else can it claim to be a reading? At another level, as may well be appreciated, there is good reason for believing that reading is a deeply complex process, in which how much the reader brings to, as opposed to observes in the text is a matter of considerable interest and not a little debate. Simply put, reading is an interpretive process in which the reader plays a crucial role in constructing the meaning of the text.[4] Thus, to suggest that one simply observes what is displayed in the text would appear to be in no small measure at odds with this.

It has to be acknowledged that this is tricky territory. Central to DA is a recognition of the importance of the reader in constructing the meaning of the text. In fact, discourse analysts often aim to study precisely this role, by observing situations in which the reading process is displayed—such as the case of creationists' reading(s) of evolution, for example! (See also Edley, 1993; Potter, 1988) Is not this a hopeless contradiction?

Yes and no. It is—but no more than that confronting any attempt to study interpretive processes. One is always offering an interpretation of an interpretation. This seems unavoidable, if it is accepted that humans engage in interpretive processes in the production of meaning. Here, everything becomes a text, everything must be read, and everything, therefore, in this sense, is a matter of interpretation. DA is, to this degree, as reliable as any other procedure or perspective one may care to propose (see the debate between Halfpenny, 1988, 1989, and Potter & McKinlay, 1989).

However, it is in some ways better. It has the great virtue of openly acknowledging its own interpretive role in the specific sense of trying to avoid advancing definitive versions of what is being studied. As Gilbert & Mulkay (1983) argued, standard practice in interpretive sociology is for the analyst to present summaries of whatever texts are under study (interview transcripts, participant observation notes, etc.), which purport to represent the frame of understanding within them, as a single, or dominant, or unifying feature. DA tries to avoid this by deliberately making a virtue of variation. Variation, as Potter & Wetherell (1994) stated, is "a lever" in the analytical process. The injunction here is: Look for the way meanings move. This might, for example, be in the way specific words or phrases are used to

---

[4]References to this central issue of concern in the humanities and social sciences are numerous. See, for example, in literary theory, Eagleton, 1983; Fish, 1980. In sociology, the issue has surfaced especially in the study of the mass media—see, for example, Ang, 1996; Barker, 1989; Hall, 1980; Morley, 1974, 1980, 1986; Stevenson, 1995; Wren-Lewis, 1983. It is also now becoming prominent in psychology—see Forrester, 1996. Also relevant is Potter, 1996, especially chapter 3.

substitute for others (and in prompting our awareness that only certain words and phrases appear at all, yet others do not). Or it might be in the way a whole stretch of discourse seems to offer one kind of understanding of something, whereas another stretch seems to offer something quite different. For example, the way something like science is understood in, or between, texts.

This provides a convenient point for an illustration. I begin by offering a very close reading of just one short section from one CSM pamphlet, in order to illustrate how DA may be done. Consider, then, the following brief passage (I have numbered the sentences).[5]

3.1   [S1] **The essential of Scientific Method, or the Inductive Process, is the discovery of facts and the drawing, or induction, from them of sound conclusions. [S2] The test of the validity of a theory is ultimately WHETHER IT IS SUPPORTED BY, OR IN ACCORD WITH THE RELEVANT FACTS. [S3] This is how the theory of evolution is to be assessed:** *not by ideas or possibilities, but by reference to well attested observations made by competent, impartial observers.*

[S4] For the conclusions to be sound, the facts must be both indisputable and sufficiently numerous to justify the deduction from them of general conclusions which may be called theories. [S5] Every other relevant factor later brought to light should be in harmony with the derived theory. [S6] Where the theory is found to fit every such fact, and moreover can be used to predict behaviour, events or discoveries, it becomes a scientific 'law'. [S7] *But if subsequent facts prove to be at variance with the theory, the latter must be modified or even discarded.* [S8] The history of science is littered with abandoned theories. (CSM Pamphlet 58, p. 1)

At first glance, this extract might seem unequivocal in advancing a view of science as involving induction; indeed, in S1, the scientific method is defined as the inductive process. How much clearer can you be?

However, despite this apparent simplicity, matters rapidly become more complicated as we move on. For one thing, this inductive process itself starts to involve complications. Already in S1, reference is made to conclusions being sound, which seems to suggest some principles of logic might be involved, as well as the simple discovery of facts. In S2, we then find that not just any old facts will do, but only the relevant ones. Then, in S3, even more factors begin to enter: Observations, we are told, must be well attested and made by competent, impartial observers.

---

[5]From this point, all extracts from creationist literature are numbered in sequence and will be referred to by their number. My own elisions are marked by dots inside square brackets as: [...]. Three dots indicate an elision within a sentence; four, an elision of a sentence or more. See also footnote 3, chapter 2.

Science, then, already, is no longer just about facts, or even facts and theories; rather, facts must be relevant. What makes a fact relevant or not? How do we decide? Moreover, facts have turned into observations. Are these the same things exactly? What does it mean to talk of observations rather than facts? These terms might be taken to allow and disallow different things. For example, whereas facts are something that might be taken as already existing regardless of whether they are witnessed or not, observations seem to imply an observer. To turn facts into observations, then, is to assign some role in the process to human involvement. And, indeed, this seems to be made explicit in the requirement that they be made by "competent, impartial observers." Also, perhaps, talk of observations gives to the account something of the technical flavor of science; so maybe this is an example of the use of scientific language to convey a sense of legitimacy? Meanwhile theories have also changed to become sound conclusions, which seems to mean being "supported by, or in accord with the relevant facts." How does a fact support a theory? How is it in accord? And is supporting the same as being in accord?

Now, the point of asking these questions is not so much to cast doubt on what the creationist is saying, as to demonstrate that one can usefully question a text in this manner. In so doing, we begin to see that there are all kinds of subtle shifts and movements going on at the surface of the text. DA contends that these movements, however trivial they may seem, are significant to the way in which meaning is constructed within the text. These movements trace out the process of negotiation of meaning, the steps and stages of the author's attempt to control his (he is male) words, so that they convey the meaning required (which is not to say the attempt necessarily meets with success!). In so doing, he draws on the resources provided by whatever appropriate discourses are available. These are employed knowledgeably in this new context in order to construct a case for one view against others—others that are often merely implied by the use of specific words and formulations as a sort of hidden backdrop or received wisdom (which itself, of course, is also open to variation when articulated).

These words are used, then, to invite the reader to accept certain kinds of connections and to follow with the text in its development of a particular way of understanding science. It is a way of understanding science that proposes considerably tight controls. The criteria encountered so far are already very strict, involving a range of factors that span the different views of science outlined previously: Empirical correspondence with (relevant) facts and/or well-attested observations; logical coherence in the form of sound conclusions; and human conventions in the demand that observers be competent (implying procedures of assessment by others).

However, the author is clearly still not satisfied with all this, as the text goes on to add yet more criteria. In S4, facts must now also be "indisputable" and "sufficiently numerous," while theories have now turned into "general conclusions" that are deduced, rather than induced. To be fair, this particular movement may be no more than the consequence of an unfortunate slip of the typewriter—and yet, perhaps not, because thereafter, in sentences S5 to S8, the text goes on to expound a view of science that seems as much in keeping with falsification as with induction. The idea that theories should be "modified or even discarded" if they do not "fit every ... fact," and that they are "used to predict behaviour" is in keeping with deduction as much as induction (see Chalmers, 1982).

The point here is not to raise questions about the internal coherence of the extract. Nor is it to question the correctness or otherwise of this representation of science. Rather, it is to highlight the variability and flexibility of meaning within it and to illustrate that we can observe usefully the way in which the terms move, where they are allowed to overlap, and where they are kept apart. To be sure, to do so can seem a tedious process; it requires paying close attention to details of terminology and formulation—or "reading for detail," as Potter & Wetherell (1994) put it. Doing this might also seem unduly pedantic, belaboring of the obvious, showing overconcern with trivial matters, or conflicting with any number of other rhetorics employed to delegitimize refined analysis (often, dare it be suggested, in the interests of advancing much grosser interpretations that find such detail most irritating when it does not easily seem to fit!). Yet, more often than not, this kind of work does pay dividends.

In this case, for example, from close examination of the text so far, clearly terms are used variably to build up a network of interleaved criteria by which a theory might be said to have failed the test of scientificity. Now, of course, one might dispute the adequacy or validity of any one or all of these criteria. What is more important to DA is to determine what their purpose is in the specific way in which they have been formulated within the text. The question is: What do they do? What work does a particular formulation—be it a word, a phrase, or other semantic order—do in its context? What does it accomplish? And in asking this kind of question, we move from the simple(!) observation of terms to a consideration of their rhetorical work.

Potter & Wetherell (1994) distinguished two aspects to this, but they are closely connected and in practice effectively indistinguishable. They are "looking for rhetorical organization," and "looking for accountability." Both refer to the manner in which a text is constructed to advance a case for one view of things as contrasted with alternatives, how the text works to persuade us toward one point of view and against another. This will be done through the use of a range of legitimizing and delegitimizing devices. One of the most central of these is the construction of factuality. This refers to the use of techniques that work to give the text the effect of describing the way things are in the real world, beyond the text itself, which it claims to be depicting to the reader. These processes will be considered in more detail in the next two chapters, but some idea of what is meant can be gained by referring again to extract 3.1.

Several points about the rhetorical organization of this extract can be made. First, it is marked by a pervasive aura of authority. Somehow, the voice of the text appears to be authoritative, giving the impression that the writer knows what he is talking about. It is a voice that lays down the law, and it is, therefore, a sure and confident reader who would dare to take issue with it. How is this achieved?

There are a number of things at work. One is the adoption of an impersonal register in the voice of the text. In one way, this shows in the absence of any modalization (Latour, 1987; Potter, 1996) to statements, in the form of qualifying terms, or statements about the statements, such as "I think that," "I believe that," or "it seems to me that,", and so on. In terms of Latour & Woolgar's (1986) hierarchy of modalization, the strongly modalized statements of this text convey a sense of concrete givenness about what is being said, such that the statements do not appear as claims made by a subject, a particular person, but as definitive, general truths.

Similarly, the text is marked by *nominalizations* (Hodge & Kress, 1993), which they define as a type of transformative process that removes any mention of (human) agency from a sentence. For example, take the clause in S1: "the discovery of facts and the drawing, or induction, from them of sound conclusions." Following Hodge & Kress, this formulation would be seen as a transformation from a root form of the following kind: "[someone] discovers facts and [someone] draws, or inducts, from them sound conclusions." Grammatically, then, both the discovery and the drawing are transformations into noun forms of original verb forms. The verb forms are taken to be the root, because they incorporate directly the agency that ultimately must be the instigator of the process (of discovering and drawing). Be that as it may,[6] the critical point to note is that nominalizations may imbue a sense of impersonality and authoritativeness.

Another feature is the way the text begins with a minimum of introduction. The extract consists of the first eight sentences of the pamphlet. Thus, there is no other prolegomenon, no other buildup to the discussion; rather, the reader is thrust directly into a series of statements that, almost literally, appear to be laying down the law to them.

Something else that might be noted are certain features of format and formal structure (Thwaites, Davis, & Mules, 1994) of presentation, in the use of boldface, italics, and upper case. These add emphasis to chosen meanings and formulations. An amusing related aside is that in the section of the pamphlet immediately following extract 3.1 appears a discussion of Darwin's theory of evolution, which emphasizes its speculative nature (a common creationist claim). This is preceded by a small drawing of Darwin's head, with the bold subheading **"Speculation"** directly above it!

These, then, are all features of the text that are designed to have the effect of persuasion to a point of view. This kind of persuasive work is so commonplace, however, that we may tend to forget that it is there, or think it too trivial to consider. Fortunately, however, DA does not stop at this level of reading; indeed, in a sense, it only really begins when other levels of argumentation are addressed.

As already suggested, extract 3.1 is advancing a range of criteria that are said to define the scientific method and the scientific status of a theory. In being set up like this at the outset of the text, they are being authorized (Smith, 1978) as grounds of assessment and made available to be mobilized as required in later sections of the pamphlet. It will come as no surprise to be told that the rest of the pamphlet is devoted to a critical dismantling of evolution, making flexible use of these criteria as and when they are deemed appropriate. These criteria, then, can be thought of as a battery of ammunition in an argumentative war. The author first equips himself with his arsenal and then proceeds into the field with all guns

---

[6]It should be pointed out that there are significant differences between the methodologies of critical linguistics and the discourse–rhetorical analysis adopted in this study. Without going into too much detail, the central differences seem to revolve around the question of transformations and the role assigned to the reader. In essence, Hodge & Kress (1993) began with a highly formal grammatical understanding, which they used in the identification and analysis of transformations and further concluded that linguistic formulations impose strong constraints on reading. However, there is reason to doubt the primacy attached to formal grammatical systems over everyday language use (Potter & Wetherell, 1987; Sharrock & Anderson, 1981). The one should not be assumed to be the root form of the other, without attention to contextual features. A similar point applies to reading, which, as already stated, is less constrained by the text than Hodge & Kress (1993) originally suggested (although they have since revised their position in this respect). For a more complete critique, see Potter (1996).

blazing—or rather, like a sniper, he picks off his targets one by one.

We have seen one example of this already, in the accusation that Darwin's theory is based on speculation. Here, a contrast structure (Mulkay, 1985; Smith, 1978) is mobilized, built around the distinction between speculation and the facts and observations incorporated into the definition of science. This is made more explicit in the next section of the text:

3.2. Arguing from the limited knowledge of his day, Charles Darwin produced a plausible speculation in the name of science [...]. While admitting a number of difficulties, especially the absence of essential scientific evidence [...] Darwin *assumed* that Evolution had in fact occurred. (Pamphlet 58, p. 1)

The specific term selected here is *evidence,* but the contrast mobilized is that between factual and speculative. There is also extra significance attached to this evidence by use of the adjective *scientific.* This term—*scientific evidence*—is commonplace in modern discourse and is given further attention in the next chapter. It harks back to the setup of the scientific, with the now-implicit meanings of relevance, attestation, impartiality, and so forth. Any one or all of these might then be utilized as grounds for the charge that Darwin's evidence is inadequate.

Similarly, one might note the presence of other terms and phrases that are also doing rhetorical work to undermine the credibility of Darwin's theory. The phrase "arguing from the limited knowledge of his day," for example, trades upon a common rhetoric of progress, with the implication that today's knowledge is superior and able to see Darwin's errors. Describing his work as an argument in itself conveys the sense that it—and, therefore, evolution itself—does not have the status of fact (another recurring creationist claim). Similarly, Darwin is described as having "admitted ... difficulties." This uses a standard trope of academic discourse to convey a sense of external veracity to the objections raised, with the added pungency of seeming to gain support from the major opponent himself. What better corroboration (Smith, 1978; Potter, 1996) can one have than to claim the support of the main opposition? Finally, the crucial matter of evolution itself is described as an assumption of Darwin's, again working to undermine any sense of (f)actuality that might be ascribed to the theory.

It is also worth noting the description of Darwin's speculation as "plausible." This is an example of *stake inoculation* (Potter, 1996) using a distancing device to appear neutral and disinterested, at least enough to acknowledge the opponent's strengths. In this way, the author deflects the potential charge of extreme bias, to appear moderate and accommodating.

However, it is notable that this appears in a context in which Darwin's ideas are being firmly located in the past. Thus, although they may once have been persuasive, the implication is they are no longer so; time is used to introduce a sufficient sense of distance such that the acknowledgement of plausibility to Darwinism does not entail commitment to it now. And this is contextualized against the earlier claim from S8, in extract 3.1, that "the history of science is littered with abandoned theories." Obviously, then, Darwinism has had its day and should be abandoned with the rest of them.

There is no need to continue this analysis further now, although much more could be said. What needs to be emphasized is that the definition of science advanced can be thought of as providing a range of means of attacking evolution to be utilized as needed. Thus, later arguments in the pamphlet make reference to the lack of corroborative evidence for evolution; unsound procedures of argument and theory construction; and evidence of fraud (such as the infamous Piltdown Man affair, another creationist favorite), indicating a distinct lack of the requisite impartiality. Each of these, it will be recognized, make more or less explicit reference to the criteria set out in the opening. Yet, these criteria themselves are drawn upon selectively and flexibly; they are made to work as required for the purposes of showing in each specific instance that evolution is in breach of the rules of science.

FURTHER EXPLOITS

I began this analysis with a discussion of the importance of variation as a methodological lever for the discourse analyst. So far, however, I have only dealt with variation within the confines of one text. A more overt level of variation is to be found when distinct texts are compared, and it is at this level where my argument with orthodox sociological representations of creationism really develops.

In extract 3.1, we saw science being presented in broadly inductionist terms. There are, however, a range of other ways of representing science to be found in my sample of pamphlets. To begin with, here is a further series of extracts, which, although a little demanding of the reader, can best be discussed comparatively.[7]

---

[7]It should be made clear that no claim to statistical representativeness in this choice of quotations is being made. From the present point of view, it is the range of discursive representations that is of main interest, rather than the frequency of their use.

3.3 The theory of evolution cannot now be presented as a *scientific* theory. The evolutionist lacks the crucial theories of heredity and development which are essential to the modern Darwinian formulations.

The scientific case for creation is—and actually always has been—consistent with all *known* principles and *observed* facts. Creationists have every reason to await future discoveries with confidence. (Pamphlet 227, p. 6)

3.4. Actually neither evolution nor creation qualifies as a scientific theory [...].

Creation has not been observed by human witnesses. Since creation would have involved unique, unrepeatable historical events, it is not subject to the experimental method. Also as a theory it is non-falsifiable seeing it is impossible to conceive an experiment that could disprove its possibility. Creation thus does not fulfil the criteria of a scientific theory, but that of course does not question its ultimate validity. [....]

Evolution theory also fails the criteria of a scientific theory. It has never been witnessed by human observers; and is not subject to the experimental method. (Pamphlet 210, p. 1)

3.5. Although it may at first seem that macroevolution is also subject to experimental analysis and possible falsification, Popper is correct in asserting that the macroevolution model is nonfalsifiable [....]

Perhaps an illustration or two of this nonfalsifiable characteristic will help. If all life were related to just one or a very few lines of descent, we should expect to find a rather complete series of fossils recording the whole transaction. As a matter of fact we find that nearly all major groups of plants and animals make their debut into the record without a fossil lineage. Do these negative results lead most macroevolutionary paleontologists to believe their model has somehow been falsified? By no means. The model is defended by a series of supporting assumptions [....]

Espousing such arguments in defense of macroevolutionism surely makes the model irrefutable but at the same time destroys its claim to scientific status. The macroevolution model becomes safe from negative evidence but has been changed from a scientific theory to a belief system beyond the pale of falsifiability. [....]

The macroevolution model [...] can be accommodated to fit widely diverse data and as such would seem quite versatile. But far from being an asset, this versatility shows that the model is not a scientific theory .... (Pamphlet 241, pp. 2–3)

3.6. Far worse for the evolution hypothesis of common ancestry are the latest data of molecular homology, amino acid and nucleotide sequence studies. Patterson said that if [Ernst] Mayr and evolutionary theory are saying anything, they must be saying that those forms more recently descended from a common ancestor have a greater similarity among their genes and gene products than those more distantly related. [....] But concerning this foundational principle of evolution, Patterson concludes: "The theory makes a prediction, we've tested it, and the prediction is falsified precisely."

Common ancestry falsified!? The first example Patterson used [...] involved [...] the amino acid sequences for the alpha hemoglobins of a viper, crocodile, and chicken. On the basis of evolutionary theory we "know" that vipers and crocodiles, two reptiles, are much more closely related to each other than either is to a bird which is presumably a much more distant relative. [....] It is the crocodile and the chicken that show the greatest similarity [...], said Patterson, with the viper and the chicken the next most similar [...], and the two reptiles with the least similarity [...]. In this particular example, the evolutionary "prediction is falsified precisely," insisted Patterson.

An isolated example? No, he continued with several more. [....]

It certainly would appear that if evolution is a scientific theory that can be falsified, then it has indeed been falsified! (Pamphlet 250, pp. 3–4)

These quotations might, at first sight, be taken as further exemplifying the same essential view of science already encountered, placing emphasis on empirical criteria of assessment. Thus, 3.3 refers to "known principles and observed facts," as well as "future discoveries," apparently adopting a discourse in which science is a progressive disclosure of reality. Similarly, 3.4 also prioritizes "observation by human witnesses" and testing by "the experimental method." It also explicitly mentions falsification in this respect. This view—of the nonfalsifiability of macroevolutionary theory—is developed further in 3.5, with specific reference to "negative results" and "negative evidence" of what "we find" in the fossil record. There is also talk of the theory fitting "diverse data." And, again, in 3.6, more "data"—indeed, "the latest"—is described and is said to have "falsified precisely" a "prediction" of evolution.

So there seems much common ground here. Certainly, taking the view that creationists adopt a simple empiricist model of science, there would not seem to be much to dispute such a view among these extracts.

However, looking more closely, and especially contextually, differences begin to become apparent. For example, although 3.3 makes much of empirical matters, it also places emphasis on matters of theory. Although there is reference to empirical concerns, this is directed immediately at creationism itself, whereas evolution, it is said, "lacks ... crucial theories." This stress only becomes clear when the wider context of argument in the pamphlet is recognized. The extract is taken from one of the most technically sophisticated pamphlets in my sample. The overall argument incorporates a critique of the theory of genetic inheritance, proposing as an alternative a theory of cortical (i.e., basically, cellular) inheritance, which it is argued is more in keeping with the evidence that "the control of any one [inherited—S. Locke] characteristic may be governed by a large number of genes, all with different roles to play" (227, p. 2). Thus, it is a discussion directed at perceived theoretical weaknesses of the theory of evolution in respect of the process of inheritance, arguing that creationism can accommodate both the genetic evidence and the possibility of cortical inheritance (presented as a "working hypothesis" [227, p. 3]), yet evolution cannot. A relative assessment of both theories is offered on both empirical and theoretical levels. Against this background, the summary assessment in 3.3 begins to take on a much greater air of sophistication. Of course, there remains reference to a world of apparently independent facts, but the view of science that is being advanced here seems to be one that understands it as involving the relation between facts and theories, and very specific aspects of modern Darwinism at that.

In contrast, 3.4 returns us to an apparently more direct view of things. The criteria of scientificity here are plain: observableness and experimental testability. Moreover, there is a further complication concerning the assessment of creationism itself. In 3.3, whatever its apparent sophistications may be, the scientific status of creationism is certain: Creationism is consistent with known principles and observed facts. We can see here a contrast structure (Mulkay, 1985; Smith, 1978) established, between consistency with principles and facts versus lack of crucial theories, in terms of which evolution fairs badly beside creation and the latter appears as definitively scientific.

In 3.4, however, neither evolution nor creation passes the test of science. The contrast here is between observableness–testability (i.e., falsifiability) versus nonobservableness–nontestability (nonfalsifiability). This is applied at the most fundamental level of the processes of change each theory purports to describe: God's creative act, on the one hand; and the gradual developmental process of evolution spread over eons of time, on the other. Both of these are beyond the human scale of existence and, therefore, beyond our ability to test.

In 3.3, then, we have a view of scientific authenticity into which creationism fits—in 3.4, a view in which it does not. This kind of situation is often pointed to by critics of creationism as evidence of their confusions and contradictions. I return to this point later.

Meanwhile, what of extract, 3.5? Here, again, further complexities seem to be involved in the assessment of the scientific authenticity of evolution. This extract is taken from a pamphlet that makes use of a distinction, drawn from noncreationists, between macroevolution and microevolution. The author proposes to reserve the second term "to designate in living organisms those changes that can be studied by field and/or laboratory procedures" (241, p. 2). Macroevolution, on the other hand, is "to cover the concept that all taxa are somehow descended from a common ancestry over vast periods of time" (241, p. 2). Thus, it is suggested that only the first of these can properly be considered scientific, because it is subject to the criteria of falsifiability specified by Popper. The second—macroevolution—however, is not.

This extract is interesting to consider, not only in relation to other extracts from the CSM pamphlets for their varying views of science, but also in relation to the charge from critics of creationism that they have a naive understanding of falsificationism (e.g., Kitcher, 1982). It is not my intention to try to defend creationists from this kind of charge, but nor should we ignore complexities when we find them. Here, for example, the author bases his (he is male) claims about macroevolution, not only on quotations from Popper and Medawar, but he also directly takes issue with attempts to defend macroevolution with "supporting assumptions" about the apparent lack of requisite fossils. The point to be made is that such supporting assumptions are the kinds of things that might be referred to by sophisticated falsificationism as part of the "bundles of hypotheses" under test (Kitcher, 1982. The term was originally Duhem's. See also Potter, 1996, pp. 22–23). The argument in this pamphlet, however, is that macroevolutionary theory is so flexible that new hypotheses can always be added to the bundle to account for missing evidence; it allows too much of what Chalmers (1982) called ad hoc modifications. In other words, it is less that the creationist holds a naive view of falsification and more that he is disputing the adequacy of a scientific theory that seems to allow such flexibility of application.8 Thus, to view the creationist simply as a naive falsificationist would be to ignore what seems to be the point of the argument.

Regardless of the validity of this argument, however, the important point is that the view of science here seems far from simple or naive. Nor, incidentally, is it easy to see in what sense the use of Popper and

---

[8]That having been said, it should also be pointed out that the creationist's use of Popper here does not appear to recognize that Popper changed his opinion of Darwinism, coming to see it as "a metaphysical research programme" (Popper, 1974a).

It would seem appropriate at this point to note that the adequacy of creationists' quoting and citing of evolutionists is sometimes called into question by their more strident critics (e.g., Cracraft, 1983; Ruse, 1982). From the present point of view, this is not an issue of great concern in itself, as my interest is in the representations themselves, not their veracity. In this respect, in fact, what is of greater relevance is the observation that, as much as critics might question the honesty of creationists, so do creationists highlight any apparent exposure of evolutionists. Charges of fakery and deception come from both sides. This symmetry of accounting is discussed further in chapter 4.

Medawar here might be taken as symbolic, as Toumey (1994) proposed. Most importantly from the present point of view, the understandings of science being drawn on seem to be multiplying.

Further variation is introduced in extract 3.6, where further apparent contrast can be observed, in this case with extract 3.5. Whereas 3.5 presents macroevolution, defined in reference to the thesis of common ancestry as nonfalsifiable, 3.6 presents a view of common ancestry as having been falsified! What might also be noted here is that common ancestry is itself defined in a relatively precise manner, involving the distribution of genes in populations of supposedly related species. As is made plain in the extract, this is viewed as a prediction, or, it might be said, a hypothesis derived from the theory of common ancestry understood in these terms. What, then, is to be made of the claim that the theory is nonfalsifiable? And what also of the claim in 3.4 that evolution is not observable and therefore not scientific?

Again, it would seem that, if one wished to find support for the view that creationists present contradictory views of science and the standing of evolution theory (or, indeed, creationism itself) in relation to it, then one need look no further. This, however, is not what is of most importance. It is in fact extremely easy to find examples of such contradictions (if they are such) in creationist texts. But then, if creationists themselves are to be believed, it is likely that a close examination of the writings of a range of evolutionists would show similar apparent contrasts. This is a common claim found in my sample of pamphlets, used by creationists as a means of undermining the credibility of evolution (but see footnote 8). Further, I have showed that it is possible to extract different, and not necessarily compatible, views of creationism from discussions of it by sociologists.

These contrasts arise, not because of shoddy thinking, ignorance, or misunderstandings, but for the simple reason that views of the nature of science do actually vary. More correctly, within the discourse of science, there are available a number of different representations of the nature of science. These representations are all potentially available as resources to be exploited for the purposes of constructing stances, arguing cases, and persuading for and against different points of view. This is what the sociologists do in their discussions of creationism, seeking to undermine its credibility as science in order to advance socially determinist explanations; and, this is what creationists do in their critique of evolution. They use the resources available from within the discourse of science, in order to undermine its credibility. And they are far from unaware of the irony of this argument; rather, this is central to their strategy. If they are able to undermine evolution on its own territory—that of science—then so much richer the victory, or, more to the point, so much more persuasive the case. This is not to say that other discourses (moral, religious, etc.) are not also employed; it is, however, to highlight the similarity of the rhetorical techniques used by the ostensibly different parties in this dispute.

However, the multiplicity of models of science utilized by creationists does not stop here. So far, I have focused only on universalist models of science. The following series of extracts provides examples of more socially localized views of science being drawn upon within CSM pamphlets:

3.7. True science involves a willingness to cast aside pre-conceived notions, the strength to stand up to peer pressure, and a commitment to follow the search for truth wherever it might lead.

[....]

Dr. Patterson's open and objective approach to the evidence seems to embody the scientific spirit we encourage in students, but so rarely see in practice among professionals. It's a sad cliche in science to say that old theories never die, only their proponents do. [....] But science has precious few examples of scientists who, because of the evidence at hand, have made paradigm shifts at the peak of their professional careers. Dr. Patterson is thus to be highly commended for his intellectual honesty and the courage to face possible ridicule from his scholarly peers. (Pamphlet 250, pp. 1, 4)

3.8. Sir Ernst warns that "Extravagant claims have been made in the field of genetics, and here the blame must be laid [...] on the scientists themselves. Actually, scientists are often just as prejudiced in their theories and as emotionally involved in the implications of their work as are other non-scientific members of society, and are unreliable in their predictions and interpretations.

[....]

He thinks "there is no reason for believing that scientists are better qualified than others to give advice in political matters. Scientists are highly skilled specialists in specific areas and obviously better qualified than anyone else to give advice within fields of their professional activity, though, even in these, different scientists may hold very different opinions." Further he states, "There is no evidence that scientists per se have any greater claim to wisdom than any other members of society; there is, in fact, a good deal of evidence to the contrary. The view that scientists are objective, dispassionate, impartial and tolerant is a myth. They are just as prejudiced and emotional as any other group of people, certainly in matters outside their professional competence, but even in their own fields of research, in relation to the views of colleagues with whom they disagree. Their power of logical thinking is also not above that of other professions." (Pamphlet 204, pp. 2–3)

3.9. Because creation involves science, many people decide to leave this aspect of their faith to the scientists. They somehow regard scientists as being unbiased and infallible. However, the truth is that scientists are no more objective than the rest of us. The atheistic scientist will fit his observations into the framework of his atheistic philosophy, while the scientist who is a Christian will fit the same observations into a different framework. [....] The question is not whether a scientist is biased or unbiased, but which bias is the best bias to be biased with. (Pamphlet 266, p. 1)

3.10. Many people believe that an acceptance of all that the Bible teaches is irreconcilable with the discoveries of science. [....] Where disagreement does exist, it is in fact the result of conflicting *interpretations* of both science and Scripture. [....]

In any consideration of the supposed clash between science and the Bible, it is necessary to distinguish observation from interpretation, or fact from theory. Natural phenomena—the facts—are undeniable, but how are they to be understood and interrelated? It would be foolish to suppose that scientists view the facts apart from any preconceived theories or conceptual frameworks. A good theory involves the construction of an abstract model which fits the facts. The model should enable predictions to be made and should be capable of disproof by means of crucial experiment.

Christians would not accept any theory or interpretative framework which denies the existence of God or conflicts with the clear statements of Scripture. They are free to construct a model of the universe, including all the scientific data, but using the Biblical framework of reference. (Pamphlet 245, p. 1)

What, I hope, is apparent from this series of extracts is the further diversity of ways of representing science to be found within CSM pamphlets, ways that refer, in one way or another, to the human and social level of science.

Thus, 3.7 presents a view that appears to owe much to the ideas of Thomas Kuhn (1970). There is talk of paradigm shifts and the cliche—originally attributable to Planck—that "old theories never die, only their proponents do." A further point can be made here, however. This extract comes from the same pamphlet as 3.6, which presents the claim that common ancestry has been falsified. Apparently, then, the author shifts between Popperian and Kuhnian views of science. However, this shift occurs through the mobilization of a contrast structure, in which science is defined initially in the form of an idealization of the individual scientist as truth-seeker, as against the sad reality of closed-mindedness within the scientific community. The contrast is used as a means of first establishing, and then exposing, an ideological self-image, and it makes use of the resources provided by Popperian and Kuhnian characterizations of science in order to do so.

Thus, from Popper, it is possible to take the idealization of the scientist as truth-seeker, boldly conjecturing where none has dared conjecture before and bravely submitting each conjecture to the merciless test of refutation (cf. Popper, 1974b). Equally, from Kuhn it is possible to take a precisely contrasting image: the scientist not as hero, but as galley-slave, chained to the oar of the prevailing paradigm, an instrument that hopelessly encumbers even as it provides the means of locomotion. Thus, the contrast between the scientist as heroic leader and as slavish follower is one provided by the discourse of science itself. In this pamphlet, these resources are employed to present a view of the scientific community as routinely failing to live up to its own ideal—apart from the odd exceptional individual, whose very exception makes the general rule plain.

Something of a similar contrast is at work also in extract 3.8, although there is also a difference in focus. Here, the primary emphasis is on matters of objectivity and subjectivity, rather more than individual and community. Thus, although a contrast is drawn with the imagery of the scientist as objective truth-seeker ("objective, dispassionate, impartial, and tolerant"), this is set against a reality of subjective, personal, and deeply committed involvement. In this case, then, it is less a matter of peer pressure enforcing conformity to the prevailing paradigm, and more a matter of individual human frailty and weakness that is highlighted. There is, however, some similarity to 3.7 in the sense conveyed of a popular myth about science being disabused.

One further feature of 3.8 is worth noting. There appears also to be a contrast at work between science and other areas of social life and thought. In particular, a contrast between science and political matters is suggested. This is part of a fuller argument developed in the pamphlet, which draws upon a contrast between the scientific and the moral. Science is said to have "no moral or ethical quality" (204, p. 1). This is reminiscent of the socio-logical argument seen before, that creationists contravene the logical limits of science when they attempt to assess it in moral or religious terms. Only here, of course, it is used by the creationist to construct a case against the involvement of scientists in politics (especially concerning military and industrial uses of technology—cf. the discussion of Merton in chapter 1) and biologists "playing God" (with respect to genetic engineering, particu-larly). The upshot of the argument is to suggest that Darwinism is at least one major plank in the spread of a "mechanistic concept" of the world, which is associated with a "naive [...] euphoric attitude to the potentialities of science" (204, p. 3) that ignores the human limitations of scientists.

Again, then, we can see similar representations of science being mobilized by creationist and sociologist alike, albeit to different ends. The two groups appear to share a discourse of science, which offers them the same basic repertoire of characterizations and representations, together with some stock arguments and techniques of application. In particular, a range of stances may be taken toward science itself, involving the mobilization of types of contrast structure with which to construct oppositions articulating around the boundary between the scientific and the nonscientific. The precise features of the particular model utilized in any instance helps to determine exactly where the boundary is drawn and, from this, what is included and excluded. It then only remains to draw out a characterization of the opposing entity—be it creationism or evolution—that shows it to have those features that place it among the excluded.

To add to the complications here, there is a further level of irony to creationist discourse, in that the repertoire of models of science also provides the means whereby creationism itself can be placed in the excluded category, outside the boundary of science, and yet still be considered superior to science. One version of this type of argument was seen in extract 3.4, drawing on a falsificationist model. In extract 3.9, however, a rather different model of science is employed, in conjunction with a similar contrast structure as seen in extract 3.8.

In extract 3.4, both creation and evolution are excluded from the boundaries of science on essentially empiricist grounds. It is not human subjectivity that excludes them. In extract 3.8, on the other hand, the contrast structure objectivity–subjectivity is used to advance a case against the authority of scientists in matters of morality. Here, it is not argued that either evolution or creationism are unscientific as such, but that scientists themselves contravene the boundaries of science because of their subjective human qualities.

In extract 3.9, a slightly different position again pertains. Here, as with extract 3.8, a contrast structure of objectivity–subjectivity is mobilized, but it is used to suggest that this applies at a basic level of conceptualization, such that, as with extract 3.4, both evolution and creation are not scientific in the sense of being "unbiased and infallible." But in contrast to extract 3.4, this exclusion is not for empirical reasons, rather it is to do with the nature of the process of conceptualization itself. Humans cannot help but be biased, because the subject is already implicated in the process of conceptualization. Thus, the atheist will interpret things in terms of an atheistic framework; a Christian in a religious framework. Note that, nevertheless, this does not leave a complete relativism, as it is left to decide "which bias is the best bias to be biased with."

This stress on interpretation is also found in extract 3.10. However, once again, there is a boundary movement. Whereas extract 3.9 appears to place both evolution and creation outside of science (at least in terms of its idealization as unbiased), in extract 3.10, creationism at least is moved back inside the boundary, which is defined using a combination of the conceptual framework interpretivism of 3.9, together with a suggestion of hypothetical–deductivism, in the talk of "predictions" and "crucial experiments." In this way, the possibility of a defense of a Bible-based science is opened up, using the lever that the interpretivist model provides. In this case, in fact, the defense is used in the pamphlet to construct a case against nonfundamentalist Christianity.

We see, then, from all this, something of the complexity of potential levels of argumentation that creationists utilize, drawing these out from the resources provided by the repertoire of models of science, with its central internal contrast between the asocial universal and the socially local, and all that goes with this (objective vs. subjective, individual vs. community, and so on). Using these resources, creationists are able to shunt both evolution and creation theories across the boundary of the scientific and nonscientific, sometimes to oppose them, sometimes to unite them, but always using the boundary to persuade against the credibility and legitimacy of evolution in the end. Thus, the apparently simple issue of science versus nonscience ends up providing a multiplicity of contrasts, just as a single beam of sunlight, when looked at from a certain perspective, is seen as a spectrum of many contrasting colors.

SUMMARY

In this chapter, I have looked in detail at the representation of science in CSM pamphlets with the aim of showing that creationists are able to draw on a range of different models of science in order to construct their case against evolutionary theory. These models are utilized selectively and flexibly, being applied in accordance with the specific argumentative end in mind, albeit within the same overarching goal of undermining the credibility of evolution.

Thus, it seems there can arise an apparent paradox: That apparently contrasting and contradictory means can be used to the same general end. As we have seen, creationists are able to utilize a range of universalist models of science, in order to advance criteria of scientific authenticity against which they claim evolutionary theory fails. Equally, however, they are able to mobilize socially grounded criteria of scientific authenticity as ways both of disputing the scientific claims of evolution and, at the same time, defending the relative merits of creationism. In effect, from the evolutionist point of view, this amounts to a no-win situation. For, if universalist criteria are advanced, creationists can either turn these same criteria back on evolution, or advance conventional criteria in contrast; but, equally, if conventional criteria are advanced, creationists can also use these same criteria to undermine evolution and build a case for creationism.

This "loop-the-loop" of arguments arises because of the internal dilemmatic of science, which itself provides the rhetorical resources, in the form of a repertoire of models of science, that enables these arguments and counterarguments to be devised. From the point of view of modern culture, then, it is not that science is either universalist or local, it is rather that it is both, which is to say that it may be discursively characterized with equal persuasiveness as either (cf. Latour, 1987; Prelli, 1989b). Consequently, we find similar rhetorical resources in this respect being utilized by both creationists and sociologists commentating on creationism.

There are further levels to these contrasts, however, yet to be considered. For one thing, although it may be the case that creationists are able to mobilize a range of models of science in order to undermine the scientific status of evolutionary theory, this still leaves them with the problem of accounting for its success. This will be considered in the next chapter. Also, there is the question of how creationists reconcile their views about science with their religious commitments, especially given that their fundamentalist brand of Christianity is opposed by many other Christian groups. This is discussed in chapter 5.

4

Why God Made Evolutionists

The last chapter considered the manner in which creationists draw on different views of the nature of science as resources that are employed flexibly to undermine the scientific status of evolution and to justify the acceptance of creationism itself as science. Two related points of somewhat different status follow. First, it seems apparent that the category of science itself functions as some kind of legitimizing device. Second, it is not enough for creationists simply to undermine the validity of evolution in this way; they must also endeavor to explain its existence as a competing account of reality. This problem arises from the primary assumption implicit in creationism that there is only one way to properly understand reality and, more fundamentally, that reality is singular rather than multiple. Hence, there is a common creationist tendency to assert a direct opposition between creation and evolution. The question then inevitably arises: How can the existence of evolution be accounted for? And, to sharpen the puzzle, how is it that so many people, both scientists and laypersons, can be convinced of the truth of evolution, if in fact reality is not like this at all?

This problem is a specific instance of a more general feature, which has been identified in other settings. Gilbert & Mulkay (1984; Mulkay & Gilbert, 1982) observed that the biochemists they studied were confronted by the same kind of difficulty. They argued that a primary assumption within the scientific community is that reality is singular; consequently, where there is dispute between scientists about natural phenomena, they have available a range of interpretive means to account for the presence of the competing version(s). Gilbert & Mulkay (1984) referred to these devices as the *contingent repertoire,* considered later.

Pollner (1987) took a much broader view. He argued that it is not just scientists who have to deal with the problem of the existence of competing versions of reality; rather, this is a general feature of social accounting. We all have to deal with competing versions of reality, as these are an endemic feature of everyday life, found, for example, in the context of courts of law, between prosecutors and defendants (Pollner, 1974). However, Pollner argued, it remains a primary assumption of what he called mundane reason that reality is singular and not multiple. Thus, there exist a range of devices within commonsense for managing the existence of competing accounts and ensuring that this assumption continues unthreatened (cf. Yearley, 1985).

A general way of thinking about what is going on when people attempt to account for different versions of reality is that they are seeking to delegitimize the versions they wish to reject (Potter, 1996; Wetherell & Potter, 1992), by undermining their validity or acceptability. Equally, however, as part and parcel of the process, they are likely also to be trying to persuade us to accept a favored account and to find ways of legitimizing this version. Thus, the process of legitimizing one version is intimately entwined with the process of delegitimizing alternatives. This follows simply enough from the nature of rhetorical processes. Arguments are always contextualized, which, at one level, means no more than that they are designed with some specific target(s) in mind. Thus, an argument will be constructed in such a way as to advance the case for one view, yet simultaneously undermining the case(s) for another view(s).

The attempt to legitimize creationism through mobilization of the category science, therefore, needs to be understood in this broader context. Thus, although it can be said that creationists attempt to legitimize their position in this way, this neither fully exhausts, nor properly articulates what they are doing. The mobilization of science needs to be seen in the fuller context of the range of legitimizing and delegitimizing rhetorics that they employ. These rhetorics will now be examined more closely. This chapter looks at the discourse of empiricism, or the construction of factuality (Potter, 1996), and the means of accounting for evolution in creationist texts. The next chapter considers their theological discourse.

THE EMPIRICIST REPERTOIRE

A key feature of Gilbert and Mulkay's (1984) analysis of the discourse of biochemists was the identification of what they called the *empiricist repertoire*. This refers to certain aspects of the way in which a sense of factuality is constructed. Potter (1996) suggested there are three basic aspects to this: first, "a grammatical form that minimizes the involvements or actions of the authors"; second, "data are treated as primary"; third, "laboratory work is characterized in a strongly conventional manner as being constrained by rules which have a clear-cut and universal application." (p. 153)

This suggests two possibilities, which seem to point in somewhat different directions. First, creationist texts could be read to try to determine if there are examples of the use of the empiricist repertoire to be found in their discourse. This could help to decide how scientific they are, and it would help to highlight how they draw upon science as a resource through the adoption of scientific language (cf. Nelkin, 1982). However, this raises the question to what extent science can be taken to be defined through the empiricist repertoire. Is the inference to be drawn from Gilbert and Mulkay's analysis that use of the empiricist repertoire is a defining feature of science and part of its institutional specificity (Potter 1996)?

This brings in the second possibility. Potter argued, following Wooffitt (1992), that it is a weakness of Gilbert and Mulkay's approach that they tended to treat the empiricist repertoire as "a seamless whole" (Potter, 1996), with the implication that all the elements only work to produce their legitimizing effect when used in conjunction with each other. In contrast, Potter suggested it is useful to unravel the different threads of the empiricist repertoire, enabling further exploration of their workings as individual fibers in the weft and warp of discourse. This opens the way for two further analytical steps: first, each thread can be sought out in other contexts; second, this in turn should cast further light on the role they play in the construction of factuality and the legitimation of accounts.

Such a view swings the argument away from issues of the definition of science back toward questions of rhetorical strategy. It might be that the conjunction of elements discerned by Gilbert and Mulkay is especially characteristic of the public discourse of science (cf. Prelli, 1989a) , but this should not lead us to think either that the nonusage of these elements (together or apart) equates with nonscientific discourse, nor that every instance of their appearance equates with science. Instead, discussion is shifted away from such definitive implications toward the analysis of the techniques of persuasion that are used in a given stretch of discourse, whatever its institutional location (albeit that this is relevant to interpreting its meaning).

The relevance of this discussion here arises first from the possibility that nonusage of the empiricist repertoire might be given as grounds for rejecting the scientific claims of creationism. This can itself, however, be viewed as a rhetorical strategy, to undermine the credibility of such claims: If it does not look like science, then it cannot be science. As has been seen, however, frequently the opposite view to this is taken of creationism—that it does indeed look and sound like science. Does this mean, then, that creationists use the full-blown empiricist repertoire? In my sample of pamphlets, it is possible to find stretches of discourse that do appear to have these characteristics. One example is detailed here (see Locke, 1994a, 1994b for further illustrations).

The following extract is taken from a pamphlet entitled, "The laying down of marine sediments—a revolutionary new perspective," which discusses, the formation of strata of sedimentary rock in the earth's geology. The pamphlet establishes at the outset a contrast between the uniformitarian or gradualist view of evolutionary theory (attributed to the 19th-century geologist, Charles Lyell) and the catastrophism of Flood geology. This extract is taken from a section under the subheading, "Laboratory research":

4.1   During the latter half of the 1980s, the present author carried out laboratory tests on sedimentation in both static and moving water. If Charles Lyell had been able to do these trials rather than hypothesizing about the present being the key to the past, he would have been forced to abandon his hundreds of millions of years on which evolution theory relies so completely. The results of our experiments provide empirical scientific refutation of the vast ages of time applied by Lyell to the geological column.

A first programme of research by the writer was followed by a second conducted by hydraulic engineers at the Institut de Mécanique des Fluides at Marseilles, France. The project was completed by a period of intensive experimentation at the State University of Colorado from 1988 to 1990.

The object of the experiments was to study lamination and internal structure of strata in continuous sedimentation both in still water and in water subject to a current. Up until this time lamination had been interpreted as a superposition, layer by layer of strata, over time periods.

Samples of laminated rock were crumbled to reduce them to the original particles of varying size that constituted the rock. The particles were sorted by sieving and the largest particles were then coloured to make them more visible. All the particles were then mixed together again and allowed to flow into a flask, first in a dry state and then into water. These simple experiments showed that the laminations reformed in the settled sediments, giving the sediments the appearance of the original rock from which it had been made. The strata were reproduced, not by deposition of a succession of layers as formerly thought, but due to the effect of segregation of the larger (coloured) particles during continuous flow. (Pamphlet 281, p. 4)

There are a number of things going on in this stretch of text. I first focus solely on the empiricist features. These are mainly concentrated in the final paragraph, but the foregoing paragraphs contain interesting features that also require consideration.

The first thing to note is the use of "passive voice constructions" (Prelli, 1989a, p. 103), such as in the sentence, "Samples of laminated rock were crumbled...," and those immediately following describing the treatment of the particles. Such passivizations, like the nominalizations discussed earlier, occlude agency, imparting an air of impersonality. This is the characteristic grammar of the empiricist repertoire. The aura of anonymity is further enhanced by the use of indirect noun substitutions for the personal pronoun earlier in the text—"the present author," and "the writer," in place of "I".

Secondly, we can see also that data are given primacy in the account, what Prelli (1989a) called "an inductive style." Thus, in the second to last sentence, it is said, "experiments showed that the laminations reformed"—it is the experiments themselves that do the showing. In characteristic empiricist style, objects and events in the world are endowed with capacities to act of themselves. Moreover, the showing has implications for human action, as we are told in the next sentence that, "strata were reproduced, not ... as formerly thought." Again, characteristic of the empiricist repertoire, experimental data imposes itself upon humans demanding things of us, such as that we think in certain ways.

We can also note the manner in which this description is established in advance in the first paragraph. Here, the results of the experiments and what they do is foreshadowed, specifically in that they "would have forced" Lyell to abandon his ideas about time, of which they "provide empirical scientific refutation." Again, the experiments themselves do things—forcefully, yet!—and what they do affects directly what people are able to do. In this way the reader is instructed as to how to read—that is make sense of—what follows. In Smith's (1978) phrase, it authorizes the version of the events to be described. The use of the empiricist repertoire here, then, is especially powerful as it places an advance construction on what is to follow.

The third element of the empiricist repertoire concerns the conventions of laboratory life. Here again, this feature is present in this extract, in the description of what was done in the laboratory. Thus, we are told that samples of rock "were crumbled," "sorted by sieving," "coloured," "mixed together," and "allowed to flow into a flask." Although there is little in this that might stand comparison with the kind of technical vocabulary typical of reports in scientific journals, nonetheless these descriptions convey a sense of standard techniques. To recognize this, we can ask what has been left out of these descriptions. If, for example, we desired to undertake a replication of these simple experiments, it is unlikely that we would know how to proceed unless we were already familiar with life in a hydromechanics laboratory. How do we crumble the rock? How small or large should the pieces be? How much rock should be used? What kind of sieve? What kind of flask? How much water? What rate of flow? And so on.

The point, then, is that, regardless of what procedures were actually used, their description is designed to convey a sense of standardization and conventional regularity. Indeed, the description of them as simple serves to enhance this (as well as doing other things). What went on in the laboratory, it is implied, was in no way unusual or unconventional; that is the sense conveyed, and this is the sense the empiricist repertoire trades on in general.

One further point on this. The second paragraph seems also to add to this aura of conventionality. In describing the research as having been partly undertaken in one laboratory and continued in a second, its standard nature is implied. How else could it be that the work could be undertaken in two different places—in two different countries, yet—unless it involved standard laboratory procedures common to both? Of course, it may have been that one laboratory contained certain crucial facilities that the other did not, but, if so, one would expect some mention of this, and it is mitigated against by the description of the experiments as simple. It is also notable that the research is said to have been undertaken by different people, "the writer," and "hydraulic engineers." This again contributes to the sense of standardization. It also lends an air of *corroboration* (Potter, 1996; Smith, 1978), through the involvement of, apparently, independent experts.

BEYOND EMPIRICISM

Stretches of discourse that combine all three elements of the empiricist repertoire in quite such a clear cut way as extract 4.1 are rare in my sample of pamphlets.[1] This is an additional reason why it is helpful here to consider them as separate features of the discourse. Thus, although the use of the empiricist repertoire in this specific case is, in part, designed to convey a sense of scientific authenticity, it is not this alone that works to accomplish

---

[1]The apparent lack of primary experimental research by creationists is sometimes used to undermine their claims to scientific status. In support of this, it is certainly the case in my sample of pamphlets that reports of experimental research undertaken by creationists themselves are rare. Nonetheless, they are not completely absent. Further, there is a certain hopeless circularity about this demand. Arguably, if creationists do not do much research, it is at least in part because they lack the material resources to do so (cf. Latour, 1987; Woolgar, 1988) and they lack the material resources because they are excluded from the orthodox scientific community!

this; other features of the text also contribute. Moreover, the text is doing other things, apart from just providing an account of some experiments.

On the issue of scientific authenticity, two further features can be noted. One refers again to the mention of laboratories. Why is it that the author has troubled to mention, in quite such detail, where the experiments occurred? Partly, as we have seen, this contributes to the sense of standard procedure. However, to achieve this, it would, in principle, be enough simply to say that the work was carried out at two or more laboratories, or perhaps to mention them more obliquely. There is a reason, though, why this is not enough here, one which pertains to the scientific credibility of creationism—or, rather perhaps, the lack of it. It has been observed that groups seeking scientific acceptance, who are largely excluded from the conventional orthodoxy, take particular pains to present their scientific credentials as fully as possible, as a means of managing the credibility gap (Collins & Pinch, 1979). This seems to be the case for creationists. It needs however to be stressed that the perception of such a gap is an outcome of textual strategies; it arises from consideration of certain constructions apparent within the texts themselves. It is not, then, a matter of asserting that such a gap actually exists for creationists; rather, it is an observation that creationist discourse is in part oriented to addressing the existence or possibility of such a gap.

What also needs to be remembered here is that the pamphlets used in this study are directed at a (partly) nonscientific audience. Given this, the emphasis placed on scientific credentials can be seen as a further effort to persuade and overcome the credibility gap. Mentioning laboratories is a way of providing entitlement (Potter, 1996; Whalen & Zimmerman, 1990), whereby the author is persuading us of his (he is male) right to speak on the matter of these experiments. Mention of his direct involvement also works in this way. Thus, the detailed listing of the laboratories helps to convey a sense of scientific legitimacy (cf. Prelli, 1989b). We are left in no doubt that these are real laboratories, with imposing titles, which are, it is to be presumed, part of the accepted institutional apparatus of the scientific community. The fact that the experiments were undertaken at these laboratories, then, helps to bolster the case for their credibility and acceptance.

Out of this, however, also arises a certain tension in the text. The prevailing voice is that of empiricism, which, in the context of the standard scientific paper, is employed to occlude the personal involvement of the actual scientists. Impersonal grammatical forms constitute a code, semiotically transforming human agency into the doings of things. In this extract, however, the author does include direct reference to himself, even though other scientists remain anonymous (appearing only as generic hydraulic engineers). This opens up a potential threat to the mythical world constructed with the empiricist repertoire, where all things that happen in laboratories seem to do so entirely of their own volition. To counter this threat, the author employs indirect self-references as noted above.

But why expose this potential threat at all? It may well be that this is a problem that arises because of the need to manage the credibility gap. The author needs to establish his own scientific credentials and to convey his right to speak about the experiments. The problem then becomes one of achieving this effectively without also undermining the empiricist myth. To do so, the author employs the device of indirect self-reference, which acts to bridge the gulf between his identity as a properly qualified actor (a scientist) and his anonymous nonagency constructed from the magical denials of empiricism.[2]

A second aspect of the attempt to convey a sense of scientificity comes in the use made of certain descriptive terms. This is especially apparent in the sentence at the end of the first paragraph, where we are told, "results of experiments provide empirical scientific refutation." The use of both adjectives here seems excessive. Although it may be common usage to describe a refutation as either empirical or scientific, finding both together is more unusual. This suggests that something else is going on here, and it seems likely that this is also connected to the effort to make the case appear more robust, by accentuating its scientific authenticity. The word empirical is already strongly associated with scientific work, especially of an experimental nature.[3] So, too, is the notion of refutation, which, in this context, conveys a sense similar to empirical. There is, then, some repetitiveness in the assertion, a sense further enhanced by the almost tautologous experiments-provide-empirical formulation.

These repetitions seem to be a way of placing stress and, as such, have an effect similar to an extreme case formulation (Pomerantz, 1986; Potter, 1996). Pomerantz (1986) suggested that one type of situation in which such formulations are likely to be used is when someone expects an "unsympathetic hearing," such as when they see themselves in a defensive or adversarial position. It is easy to see why this might apply to creationists.

---

[2]Another aspect of this arises in the creationist habit of detailing the qualifications and institutional standing of those to whom they refer, including themselves. In many instances, this can be understood as an attempt to place whatever discussion is under way on a properly scientific footing (for discussion of this term, see Potter, 1996), thereby again addressing the perceived credibility gap.

[3]Chambers defined empirical as: "resting on trial or experiment; known or knowing only by experience."

Frequently, they present themselves in an adversarial position, especially through the construction of a polar opposition, "zero-sum relationship" (Gieryn et al., 1985; Taylor, 1996) between creation and evolution. Often, also, they present themselves as in something of a standoff with wider social (not to mention supernatural) forces. Their stance is also often defensive, especially in respect of their relationship with the scientific community. This is again something that follows from the address to a potential credibility gap. More than this, however, creationists often present themselves as the defenders not just of the truth of the world, but also of the truth of the Word, that is, the true Bible-based faith of Christianity.

Extreme case formulations, then, frequently occur in situations where people feel themselves under pressure to persuade against prevailing convictions to the contrary, or confronted by a deal of skepticism. More usually, however, such formulations are associated with the use of endpoint quantifications such as all, every, only, sole, and so on. Although creationists do engage in such "quantification rhetoric" (Potter, Wetherell, & Chitty, 1991), in this case, a similar effect seems to be achieved through repetition to the point of extreme redundancy, as a way of attempting to dispel any doubts we might be harboring over the genuineness of the scientific authenticity of what is going on.

Thus, this passage employs a number of other techniques, apart from those of the empiricist repertoire alone, to bolster the case for its scientificity. Previously, I also said that there are other things going on here. We have seen something of this in the author's identity work that is present, which, although linked to the use of the empiricist repertoire, also draws on formulations that are independent of this set of interpretive devices.

There is one other feature to consider before moving on. I have argued throughout that discourse is constructed with rhetorical intent. This is certainly true of extract 4.1. Of course, everything that has been said about this passage so far has attempted to describe particular aspects of its rhetorical construction. However, much of this has, in a sense, been concerned with talking around the text, in that, although features of its content have been picked out for scrutiny, as yet little, if any, reference has been made to what seems to be its direct target of attack. What also needs to be considered, then, is how this argumentative context is itself displayed and constructed.

The general point here is about the nature of argumentative discourse. Such discourse defines its target of attack during the process of argumentation, as an opposing position, against which the points being advanced are to be read and understood. The opponent appears as a negative image, or as a lock to be opened, with the rhetorical key provided. This lock and key are made of a peculiarly malleable substance, one that can be continuously shaped and reshaped over the course of their engagement. Every argument, therefore, is part of a coupling that presents both itself and its opponent in a suitably stylized fashion; they are made to fit.

This can be seen in extract 4.1. Take, for example, the middle sentence of the first paragraph, beginning, "If Charles Lyell...." This one sentence does an enormous amount of work in telling us what is going on in the text. I have already said that it serves as an authorization of the version of events to be presented, instructing us in their meaning. It is this meaning that I now want to focus on a little more closely. There are three features of importance to note.

First, there is an important contrast structure in the first phrase that establishes the terms of debate, between doing trials and hypothesizing. This draws on a background of ideas about science, exploiting a tension central to the modern scientific worldview, between reality in itself and our ideas about it. In this context, the tension is used to advance a normative evaluation of Lyell's actions, one that invites us to prefer trials—experimental testing against reality—to hypotheses—speculations or conjectures about reality.

However, although this contrast already seems to trade on a cultural background of expectations in which, as a matter of course, experiment is more highly valued than ideas, it is recalled from chapter 3 that creationists sometimes mobilize other criteria, in which facts appear second best to ideas, because, for example, of the need for interpretation. We should not presume, then, that the contrast automatically favors one side (reality) over the other (ideas). There ought to be other things at work in this specific instance designed to persuade us to favoring trials. Indeed, there are two things in particular.

One is the "if ... rather than" formulation that provides the scales between the weighing pans of trials and hypothesizing and helps determine that the swing is in favor of the former. If ... rather than is used precisely as a means of constructing an evaluative choice, in this case, favoring trials over hypothesizing.

A second feature makes this much more explicit and provides far more detail about what is wrong with Lyell. This is the second clause of the sentence, which informs us of the implications of the trials, that is, that Lyell would have been forced to abandon his ideas about time. With this, we learn in greater detail about the issues at stake; in effect, this clause constructs the trials in question as a crucial experiment in deciding between views of the world. Although not explicitly stated, there is a second contrast introduced at this point, between "hundreds of millions of years" and some other period of time. From earlier indications in the pamphlet, and also from a general knowledge of creationism (see chapter 2), it is known that this other period of time is much shorter than hundreds of millions of years. Creationist opinion varies considerably on this point, but many British creationists are "Young Earthers," claiming that the earth (and everything else) was created only a few thousand years ago, in accordance with what is presented as a literal reading of the Bible. It is no surprise, then, that the orthodox scientific view of the age of the earth, in the region of 4,500 million years, comes under considerable attack. Similarly, Lyell's uniformitarianism is also a major target. Thus, what is being established here is that trials concerning the rate of sedimentation of rocks are crucial to the choice between Lyellian uniformitarianism and creationist catastrophism.

It is also worth noting that this critique is picked up later in the text in the description of the experiments as simple. In the light of the critical context, the implication is that Lyell is even more guilty in his hypothesizing, because the trials he might have undertaken are not only crucial, but very easy. This is an example of what Potter (1996) called *minimization*. It is used here to attempt to further embarrass the long dead geologist.

The argumentative context in this sentence does not, however, stop with the criticism of Lyell, as is made plain by the final clause, "on which evolution theory rests so completely." This does not merely complete the sentence; it completes also a chain of connections that the creationist encourages us to make. In effect, we are told, not merely that these trials are crucial for the determination of the age of the earth, but for the theory of evolution itself, which "relies so completely" on Lyell's "hundreds of millions of years." Indeed, it is this that makes them trials.

It is this chain of connections that constitutes the full argumentative context for this pamphlet. For this chain to be made, however, the links need to be forged in a specific way. Not only does Lyell's work need to be constructed as hypothesizing, but it also needs to be framed in uniformitarian terms, with this taken to mean certain specific things about the process and causes of sedimentation. This, in turn, must be taken to mean certain specific things about the length of time involved and then related to the theory of evolution in a quite specific way. Each link has, then, to be designed to suit. It also has to stretch out in the other direction toward the trials themselves, which need to be presented as suitably crucial; this is achieved, as we have seen, by the emphasis on their quality as a refutation.

Of course, not all of this work is made fully explicit in this one sentence, but it is implicit in the way the sentence is formulated; unpacking it shows how the chain of connections has been forged and the argument constructed. Thus, what is presented as a simple matter of experimental trials, and of simple empirical fact, can be seen as the outcome of a relatively complex construction of connections, tightly packed into a single sentence, only as an outcome of which can these trials be made to appear as trials.

One further thing should be stressed. In endeavoring to bring out the argumentative context in this way, I do not claim to be making a grand discovery, or to be revealing hidden mysteries and secrets. Nothing mysterious is happening here. No doubt much, if not all, of what I have said about extract 4.1 will seem quite obvious. Yet, it is precisely this obviousness that I am seeking to consider. Because it is obvious, there is every chance that it will not be seen as needing attention. All the more reason, then, to study the obvious to better understand how obviousness is achieved. Further, the assumption here is that the kind of constructive work analyzed previously is a feature, not just of creationists' discourse, but of all discourse (though not in these particulars of course). No discourse is pure; it always carries the baggage of past and present arguments along. If we start with this assumption, at least we will be prepared to begin looking for the baggage, even if at first it is not so obvious.

CONSTRUCTING INVARIANCE

Moving along, the kind of description of experimental work found in extract 4.1 is relatively unusual in my sample of pamphlets. More common is for the various elements of the empiricist repertoire to appear separately, as part of contexts of argument where other kinds of work are also significant. This is the major point of the previous discussion: Empiricist discourse is only part of creationist discourse. Consequently, the analysis is better served by a broader focus, one that is directed toward the processes of legitimation and delegitimation that are utilized. Of special interest in the prior case is that many of these other processes can also be seen at work in a stretch of discourse that is ostensibly empiricist in form. In other words, empiricist discourse itself can be used to accomplish diverse tasks in a given passage, being couched in relation to other discursive features that are designed to accomplish other things beyond empiricist work as such.

To consider this further, I consider one example of a more common form of presentation in the pamphlets: natural history, consisting of descriptions of particular zoological features of the natural world. In a sense, these descriptions can be thought of as substituting for experimentation. Although creationists may not undertake—or at least report on—much laboratory work, they seem to have a great interest in observing nature's own laboratory, so to speak, in the form of animal behavior, plant functioning, and the workings of ecological systems. These descriptions, then make for an interesting comparison to the features of the empiricist repertoire. In particular, they appear to be designed to achieve a similar effect, but do so through modification of the discursive elements involved.

The following extract is taken from a pamphlet entitled "The phenomena of nest-making inexplicable by the evolution-mechanism." The pamphlet has a lengthy embedded quotation, which I have reproduced in full, as it is central to the analysis I am building.

4.2   The nest of an Indian weaver bird is a beautifully plaited structure, in shape like a chemist's retort—the entrance being from below through the tube of the "retort." It hangs from one of the lower branches of a tree, and is composed of thin strips torn from blades of long elephant grass. Douglas Dewar, who long ago wrote a paper on "The Nesting Habits of the Baya" (*Journal of the Bombay Nat. Hist. Society*, 1909), tells us how it is built.

"Having selected a branch from which to suspend the nest, the bird flies to a clump of such grass, alights on one of the upright blades and makes with its beak a notch low down in a neighbouring blade, grips the edge of this above the notch, pulls or rather jerks its beak away ans strips off a thin strand of the blade. Holding this in its beak it strips off a second and third strip and flies off with these. The bird then twines these one by one round the branch from which the nest will hang. Owing to the silicon in the grass the latter does not slip when twisted round anything. The fibres next collected are plaited into those attached to the branch until a rope about 4 inches long has been plaited. This is then extended by the addition of further material into a bell-shape structure, which will form the roof of the nest. The next step is to plait a loop across the base of the bell which now has the appearance of an inverted basket with a handle. When the loop is completed the hen collects no more material but remains perched on the loop and helps the cock to plait in the material—she working from the perch, and he from the outside of the nest. The next step is to close up one side of the loop to form a receptacle for the eggs. The other side is left open, but is prolonged downwards to form a tubular passage about 6 inches long and 2 in diameter. Thus the entrance to the nest is from below."

[....]

According to the evolution theory, birds evolved from reptiles, which laid their eggs on the bare ground; so the earliest birds must have done likewise. [....] ...[I]f an evolutionist is sure that the step-by-step idea also holds good for **making constructions**, he has to speculate and to invent possible evolutionary series of steps for our nest [...]. [....]

> Well, there are cases where it is impossible to invent little steps—and our nest is one. Certain activities must be completed straight away, and in the right way; otherwise, the result of the activity is useless nonsense, fatal to the species. [....] The making of a suspended nest necessitates from the start certain acts that have to be done—quite different from those needed for making a normal nest. **The idea that everything can be achieved by evolutionary steps, and that "links" can connect construction to construction, is wrong. For you either put leaves, etc., together, or you do not.** Just as, if you want to have a steam-engine with a piston, certain things have to be done at once. And if you want to have a steam-turbine it is not possible to start with a piston and then get a turbine step by step. For technical reasons it is just not possible to transform a piston step by step into a turbine and produce a construction that is workable at each successive stage. (Pamphlet 43, pp. 1–2; spellings as original)

There are, of course, many things going on in this passage. Two features are relevant to the issues concerning the empiricist repertoire. The first is the way the description of the weaver bird's nest-making activity is put together; the second concerns the analogy with steam engines.

Building Nests. The description of nest-making, contained in the quotation from Dewar, has two particular, though related, features of interest.

The first is what I call, the no bird–every bird quality of the description. The point here is that the description employs a formulation in which, although it seems to be the activities of one particular Indian weaver bird described, this particular bird stands for any and all Indian weaver birds. Thus, although the description refers to "the bird"—one particular bird—and to 'its' behavior, the sense imparted is of a description of Indian weaver bird behavior in general. This could, therefore, be any, or every, such bird. It is not just about what one particular bird happened to do on one particular occasion, but what Indian weaver birds do as a matter of course.

This quality of the description has an interesting similarity with the element of the empiricist repertoire pertaining to the characterization of laboratory life. It invests the description with a similar sense of universalism. This is not suggesting that the activity of the bird took place in a laboratory; it might have occurred in a similar place, such as an aviary, but we are told nothing specific either way. Regardless, the point is that, although conventions of laboratory or related kind of life are not mentioned, the description is nonetheless not conventionless. Assuming it does refer to wildlife activity in the field, then we have also to assume that the observations of the bird were made employing standard ornithological techniques. We can imagine, then, a suitably attired Dewar, equipped with the appropriate means of bird-watching—binoculars, maps, note paper, sketchbooks, pens and pen-cils—out in the field, observing. Such equipment and its manner of usage are all shaped by conventions established in Western culture, reflecting the general project of zoological observation, itself framed within the terms of the scientific tradition. Bird-watching is as much a socially organized and encultured activity as any other.

In this respect, at least, then, there is some similarity with the empiricist repertoire. However, the parallel goes a stage further through the no bird–every bird quality of the description. This, too, is a matter of conven-tion. All scientific description, not just zoology, is presented in such a manner as to bridge the particular and the general. A central assumption of Western science is that nature works in accordance with fundamental regularities, which are accessible through observation of recurring patterns. Thus, particular occurrences of particular events by particular scientists in particular laboratories are allowed to stand for general regularities, appli-cable, ultimately, in all situations, other things being equal.[4]

Essentially the same rule applies to field observation. Thus, the descrip-tion is couched in terms that are, essentially, identical to those of the empiricist repertoire, only rather than describe experiments with the imper-sonal voice of the past participle, they are described with an active partici-pant—the bird—but one which has the quality of being no bird in particular, and every bird in general. To put this round the other way, the use of the no bird–every bird formulation is a way of overcoming the same problem as that confronted in the description of laboratory work. Just as laboratory work must be made to appear universally applicable, despite consisting of particular activities by particular actors, so, too, must field observations made by particular observers at particular times in particular places, be made to seem generally applicable. The formulation no bird–every bird is a way of managing this problem.

However, in this specific instance, it does a further job for the creationist. The quality of universalism imparted can be used also to convey a sense of timeless, invariant continuity, of the constancy of nature. That this is important to the creationist argument can be seen from the paragraphs that follow the description of nest-making. What is emphasized here is that the bird's activity cannot have been produced by processes of gradual change and, more especially, that it must always have existed in the form that it is now found.

---

[4]This is not to deny the significance of replications, although this activity is by no means as straightforward as it is often presented (see, e.g., Collins & Pinch, 1993). The point is more Humean. Regardless of how many replications are undertaken, we still only have particular events occurring on particular occasions (cf. Chalmers, 1982).

Note especially the emboldened sentence, "For you either put leaves etc., together, or you do not." In other words, given that it is evidently the case that Indian weaver birds do indeed put leaves together now, in the present day, then they must always have done so. The implication is, then, that they came into being already as fully capable nest-makers and have stayed this way ever since, otherwise the species would not have been able to reproduce.

The no bird–every bird formulation itself plays a part in conveying such a view. In imparting a sense of universalism, it also has the effect of suggesting eternal constancy and, most importantly, invariability. Implicit in the formulation is a sense that the nest-building described is performed in exactly the same way, every time, by any given Indian weaver bird; it is an instinctive ritual, played out again and again, with no detour or difference in any particular. The same steps are followed in the same order precisely.

This brings in the second feature of the description; its step-by-step construction. Although this is only made explicit from just over halfway through the description (with the sentence beginning, "The next step...," repeated again two sentences later), it is nonetheless characteristic of the whole. It reads like an instruction manual for building such a nest. This contributes to the sense of unvarying continuity. It is as though every weaver bird follows this precise set of steps in every case of nest building. Thus, the description itself functions as an argument against the notion of a developmental process, especially of a gradualist kind as presented here. Such gradual change implies many, many minor variations. But the invariant nest-making behavior of the Indian weaver no bird–every bird in itself makes this seem implausible.

Metaphorical Technicalities. The other feature of this text to focus on is the few sentences at the end referring to steam engines. Here an analogy is drawn between birds' nests and steam engines, with the intention of showing why it is not possible for nests to have developed in a step-by-step manner. The crux of the metaphor, where its persuasive work is centered, is in the last sentence, which can be rewritten as:

> For technical reasons it is just not possible to tranform [X]
> step-by-step into [Y] and produce a construction that is work-
> able at each successive stage.

where X and Y stand for the particular things between which the analogy is being drawn, thereby making them substitutable, such that

X = pistons or normal nests;
and
Y = turbines or suspended nests.

A general point about metaphors is that the analogies they draw are selective (Antaki, 1994; Goatly, 1997; Prelli, 1989a). Metaphors always highlight some features of the thing in question and downplay or simply ignore others. Because of this, they are rhetorical: They do persuasive work by virtue of drawing our attention to some things over others. What, then, is the persuasive work being done by the creationist in this instance?

It helps initially to think of this metaphor as an example of a more common analogical form of modern thinking, between organic and mechanical systems. Classically, this has been employed reductively: Thus, from Galileo and Kepler to Newton, the analogy of the celestial mechanism was developed, likening the movements of the heavens to the workings of the mechanical instruments being developed in Europe at the time (Koestler, 1964; Yates, 1972); and, Descartes likened animals and human bodies to machines, helping to encourage the development of the modern "biomedical model" (Nettleton, 1995; Turner, 1983). In these instances, the metaphor is used to reduce the complexity of the organic systems involved, likening them to the technical workings of machinery, thereby encouraging the view that they can be taken apart to find out how they tick. The advantage of the analogy is that it encourages a focus on material operations, allowing other possible dimensions of existence to be ignored or even denied altogether.

Interestingly, however, this does not seem to be what is happening in the current instance. Here, rather than drawing an analogy between the organic and the mechanical in order to reduce the complexity of the former and to enable its piecemeal treatment, the concern seems to be to enhance the sense of its complexity and discourage breaking down its organicism. We cannot get from a piston to a turbine in piecemeal fashion, because these systems come each as a complete whole and certain things have to be done at once.

What the metaphor seems to be doing, then, is drawing on a sense that engines are complex devices that consist of numerous interdependent parts, each with its allotted task and all necessary for the effective operation of the machine. It is the machine as system that is highlighted. The folk wisdom of this stems from the common, everyday perception of engines as things that often do not work properly and, when they do break down, require the attendance of specially qualified people who possess the requisite knowledge of the machine's arcane inner workings. It trades, in other words, on a rhetoric of science and technology as the province of mysterious expert others.

It is important to recognize that this rhetoric may be persuasive, regardless of the extent to which it is reflective of the actual state of affairs regarding the knowledge of such machines within society. The fact (or not) of ignorance of engines bears no necessary relation to the effectiveness of a rhetorical strategy that trades on such ignorance. Crucial here is the use of the phrase, "for technical reasons." This is a commonplace device used to make a display of expert knowledge, with the intention of putting an end to further discussion without need for more detailed explanation. It implies greater expertise, but is actually a gamble on ignorance, which may be working to mask one's own ignorance, as much as exploit that of others. It makes a claim to greater knowledge that is there, but which cannot be brought to hand in the current context, and so attempts to close down the need actually to demonstrate such knowledge. In short, it mystifies.

Note also that the mystification works on both sides of the metaphor, that is, it applies to the nests as much as the steam engines. We are told that it is not possible to get from a normal nest (whatever that is) to a suspended nest in a step-by-step manner that would enable the birds involved to keep reproducing successfully, but this also trades on ignorance of the technical processes that might be involved in such a transition.

In this respect, we can also note the relative *vagueness* (Potter, 1996) of this talk of nests and engines—which the phrase "for technical reasons" does much to entitle—in contrast to the relative detail of the description of the bird's behavior. The vagueness of terms such as normal nest, and the mention of pistons and turbines with no detail given of what such things might actually be, itself works to justify the claim that we cannot transform one into another; and the detailed description of the bird's behavior adds to our sense of the subtle complexity and interdependence of the final nest.

There is one further task that the metaphor accomplishes. In investing the organic world with this sense of complexity, the metaphor serves to recharge nature with a sense of mystery that can be described as holistic ("certain things have to be done at once"). This sense of mystery is the kind of thing the rhetoric of rationalization maintains is absent from the modern disenchanted outlook, and to which creationism itself is often seen as a response. The metaphor here, however, suggests a somewhat different situation. If it were a matter of reacting against disenchantment, it seems unlikely that this kind of comparison with machines would be made, or that an organic system would be referred to, analogically, in technical terms. Such terms would be rejected. The fact that they are not suggests something more subtle is going on. Although it does seem that an alternative to the disenchanted worldview is being advanced, the language of disenchantment is itself being employed as an instrument of persuasion to this end. In this case, the sense of mystification that may be invested in expertise, by virtue of specialization, is itself used to reinvest nature with mystery. In a sense, the "wow factor" of science becomes the "wow factor" of nature itself and, thereby, imbues nature with the very enchantment that, according to rationalization, science is supposed to have dispelled.

This kind of use of science is characteristic of creationism and is indicative of the syncretism between the languages of science and religion that creationists construct. This is discussed in more detail in the next chapter. For the moment, let us take further steps into this alternative depiction of nature and enter the land of design.

DESIGNING DESIGN

One prevailing feature of creationist fact construction concerns the representation of the natural world as intentionally designed. This consists of a repertoire of zoological stories, plus a range of descriptive terms, used to convey a sense of profound, unknowable mystery and strangeness. Given the variety of stories and devices employed, the best that can be done here is to concentrate on the main features. This can be done through close consideration of one extract. This passage is taken from a pamphlet entitled, "Teleology: Purpose Everywhere," which gives various evidences of teleology in different aspects of the physical world. I have reproduced one nearly complete section, focusing on biological matters.

4.3. "THE BIOLOGICAL FIELD. The fluids of the animal bodies all carry the signs of design. Blood, the carrier of the body, containing all the substances needed for its upbuilding and activities, as well as those to be eliminated, is wonderfully contrived. So too is the four-valve pump which controls its flow.

The miracle of the egg, with its peculiarly strong shell, is the story of two liquids and two cells which, without more than a steady application of gentle heat, can produce the new life of the creature which lays it. In all species the reproductive process is a miracle, beyond the conception of man's mind or remote chance.

Milk too is a marvel, eminently suitable for its purpose of feeding the young. Wonderfully secreted and containing all the substances necessary for the proper development of the offspring, it also provides a valuable food in many cases, for other creatures including man.

The alimentary canal with its attached glands and blood vessels is much more complicated, intricate and remarkable than the most sophisticated industrial plant. Essentially just one main tube, it is more than a factory producing a variety of products. Through a wise Designer, it, without apparent conscious direction, absorbs food, sorts, processes and allocates it, eliminating waste and keeping itself in constant repair. It uses and produces most complex chemicals in its digestive actions, which may be described in part, but which are neither understood nor reproduced in the laboratories of the most advanced scientists.

[....]

Numerous well-known structures of familiar creatures all point to design, rather than chance of some gradual development. The intricate organisation of the ant hill is one example. Similar are the wasp's nest and the beehive with the regular hexagonal cells. The queens, drones, workers and their particular functions suggest a Super Organiser in at the beginning. The spiders with webs produced perfectly from the first, indicate something built-in, and that needs a Builder. Birds' nests of the many species are similar structures. Who too gave the birds their light strong feathers for flight and bones of minimal weight? Man's most intelligent efforts in this form of transport are extremely clumsy in comparison.

The fine formation and beauty of diatoms; of insects, especially butterflies and moths; of plants, particularly in their flowers and seeds, are indicative of the quality of the Designing Hand behind all. The Lord reminded men that Solomon in all his glory was not arrayed like one of these. (Matthew 6, 28–29)." (Pamphlet 177, pp. 3–4)

Once again, there are many things occurring in this passage, but I confine my discussion only to the recurring use of certain kinds of descriptions to refer to the phenomena of nature.

As was the case before, nature is presented as complex. This is seen here especially in the description of the alimentary canal as "much more complicated, intricate and remarkable than a factory." The comparison with technology is again used to convey the sense of complexity, only here this is enhanced even further: It is not enough to liken the alimentary canal to a factory, in the way a nest was likened to an engine; rather, the technology comes off a poor second, as it does also in the later comparison between feathers and human flight technology. Similarly, the sense of complexity is enhanced by comparison with the paucity of knowledge in "the laboratories of the most advanced scientists." This also acts to push the comparison a step closer toward the mysterious. It is not just that the world is complicated; it is actually beyond our understanding ("beyond the conception of man"). This is conveyed especially by the frequent use of descriptions that emphasize wonderment: "wonderfully contrived"; "a marvel"; "wonderfully secreted"; "remarkable"; and of "fine formation and beauty." We might ask, however, why these terms are so common? What does this continuous stress do?

It seems to be doing three things in particular. First, terms such as these are widely employed within modern society in the description of the natural world. It is not just creationists who present the workings of nature as marvellous and remarkable; rather this is a common feature of a more general discourse within modernity. Thus, in adopting these terms, the creationist is drawing on a widely available repertoire of representations of nature and, in so doing, adopting a stance that is both recognizable and commonly accepted. The stance, in other words, appears perfectly normal. The reaction of wonderment to the world is a legitimate one and helps, then, to legitimize the further steps in the argument.

This is the second thing these terms accomplish: They involve a description of what the world is like, and they include an instruction as to how we should regard it. This is true of any description. Descriptions are not neutral; they intend a perspective. Thus, describing something as wonderful or marvellous intends a perspective in which the thing in question is to be wondered or marvelled at. In so doing, in this instance, these terms modify the view of the complexity of nature. Something that is complex (to the point of unknowability) might for that very reason be considered profoundly frightening—and this is also a common modern representation of nature. In presenting nature as wonderful, however, the complexity becomes less frightening and more reassuring, something that can be regarded contemplatively—a marvel rather than a monster. This is also helped by the stress on nature's beauty. Uniting these two senses of wonder, then, conveys the sense that it is perfectly natural (i.e., normal) to regard nature in a beneficent light.

The third thing these descriptions do is provide a link between beneficent nature and the sacred. Particularly significant here is the use of the term *miracle* to describe natural things and events ("the miracle of the egg"; "the reproductive process is a miracle"). Chambers defines miracle as: "a supernatural event: hyperbolically, a marvel, a wonder." Thus, the term epitomizes the sense of wonder, but adds to this a quite specific spin. First, it reaches beyond nature to the supernatural; second, connotatively, it connects specifically to Christianity and the biblical use of "miracle" to refer to Jesus' actions. Thus, the miracles of nature connect us directly to the divine; therefore, our common sense of beneficent wonder at nature is directly attributable to God's good bounty. These descriptions, then, work for the creationist in normalizing the sense that the observation of nature leads directly, and naturally (in all senses), to God.

It is also worth noting that the term miracle is used here specifically to describe aspects of reproduction. This adds in the further sense that it is not just that nature in general leads to God, but specifically, that life and the means of bringing it about is sacred.

That nature leads to God is made explicit in the recurring empiricist formulations referencing purpose and design: "fluids ... carry the signs of design"; "structures of familiar creatures all point to design"; spiders' "perfect" webs "indicate something built-in, and that needs a Builder"; the "fine formation and beauty" of creatures "are indicative of the quality of the Designing Hand behind all." God Himself appears in a number of indirect references, as the "wise Designer," "Super Organiser," "Builder," and "Designing Hand." These terms seem to have been chosen for their reference to a guiding intelligence; and we know, despite their obliquity, that they are referring to God, because of the characteristic first-letter capitalizations. The basic point being, then, that God is there for us to see, if we but open our eyes to look.

This way of describing the world is an example of what has been called *ontological gerrymandering* (Potter, 1996; Woolgar & Pawluch, 1985). Potter (1996) defined this as a discursive process in which "one realm of entities is constituted in ... [a] description while another is avoided" (p. 184). This is the case here, where the realm under construction is one of smooth and integrated natural functioning. In this realm, nature works properly, orderly, beneficently and harmoniously. Correspondingly, what is being avoided is a realm where natural things do not seem to go to design. In extract 4.3, there is no explicit contrast drawn that might help to bring this out more clearly, and it is perhaps an indication of the strength of the commonplace view of nature as a harmonious balance that counterexamples do not so readily present themselves. Nonetheless, it is possible to suggest where an opposing stress might be given.

Take, for example, the sentence describing the alimentary canal: "without apparent conscious direction, [it] absorbs food, sorts, processes and allocates it, eliminating waste and keeping itself in constant repair." This is all very positive. Although it may be the case that the alimentary canal does, in fact, do these things, it might also be suggested that often it does not do them as successfully as we might like. Where are the indigestion, ulcers, and stomach cancers in this description, we might wonder? And how do they fit into the "wise Design"?

The point here is not to suggest that the opposing description is more correct than the creationist's own, nor that a correct description is one that incorporates illness and malfunctioning as much as healthy operation. Rather, it is to show that the description of nature as displaying design is a constructed one and is put together to achieve specific kinds of ends; in this case, to persuade us that God, the Designer, is visibly present in and accessible through the phenomena of the natural world.

This temporarily completes what I want to say about how creationists' construct the factuality of the world and endeavor to legitimize their version of it. Next to consider is the counterpoint to this empiricist rhetoric. Initially, I said that creationists are confronted by the problem of how to account for the apparent success of evolution. As we have seen, they construct a version of the world in which it clearly displays signs that point toward God and creation, rather than evolution. If this is the case, however, they must also explain why it is that the theory of evolution persists. If the world visibly displays creation, why do people continue to hold to the truth of evolution?

WAYS OF BEING IN ERROR

This brings us back to the territory mapped out by Gilbert & Mulkay (1984) in their discussion of scientists' discourse. They referred to what they call the "asymmetry" of biochemists' accounting procedures: When discussing their own work, the biochemists tended to adopt empiricist rhetoric; but, when discussing the work of opposing scientists, they tended to adopt the quite different rhetoric of the contingent repertoire. The contingent repertoire "is in direct opposition to … the empiricist repertoire in that it enables speakers to depict professional actions and beliefs as being significantly influenced by variable factors outside the realm of empirical biochemical phenomena." (p. 57) This, then, provides the resources whereby scientists are able to account for the discrepancy between their own, empiricistically rendered, versions of reality and those of their colleagues who seem to see the world differently.

The lexicon of the contingent repertoire includes a range of tropes referring to subjective states, personal interests, and so on, as in the following list from Gilbert & Mulkay (1984): "prejudice, pig-headedness, strong personality, subjective bias, emotional involvement, naivety, sheer stupidity, thinking in a woolly fashion, fear of losing grants, threats to status and so on." (pp. 68–69) Other examples given by Gilbert & Mulkay included: interpreting everything to suit particular theories; bending the data; avoiding awkward questions; being misled by publications not subjected to proper refereeing; irrationality; having too much invested in a theory to give it up; and even being in America!

Many of these terms and phrases are also likely to be found in other, nonscientific contexts. Accusations of this kind might well be made in everyday situations as ways of undermining or rejecting views that are different from our own or which seem to present a challenge to our own experiences. Some may seem more specifically appropriate to scientific contexts (such as fear of losing grants, or being misled by poorly refereed papers), but many are likely to be generally available and more widely used. Further, even those with apparently specific application may find congruent forms adapted to other contexts (such as, for example, being misled, not by poorly refereed papers, but by low quality newspapers).

This commonality of usage arises because these terms function as ways of undermining credibility through deflecting attention to the involvement of extraneous factors and especially towards the attribution of interests, indicating some sort of stake (Edwards & Potter, 1992; Potter, 1996) in the outcome. This is a very powerful way of undermining an opponent's credibility and the legitimacy of their claims. Such a charge can always be made for the simple reason that any claim made about the way the world is, inevitably is a claim made by a particular person. When someone makes a claim about the way the world is, they claim to speak beyond themselves, free from their own subjective condition and social context. Yet, they remain a particular subject. Hence, they always remain open to the potential charge that they are speaking merely for themselves or those they might represent. Thus, a basic and powerful means of raising doubt about a claim is to point to the presence of personal or social factors.

This harks back to the central argument of this book. I have argued that there is a dilemma within science between the universal and the local: Science claims to speak a truth for all times and places, but it is a product of a specific social and cultural context. This provides the basic ground on which a critique of science can be constructed: Its general claims can be said to be merely particular. Further, the basic resources for constructing such a case can be drawn from the repertoire of devices commonly available within the wider culture as means of identifying and accounting for error.

This can be seen in creationist discourse. Creationists employ a wide range of means of error identification and accounting in their efforts to undermine evolution. Many of these are identical to items in the lexicon of contingencies given by Gilbert & Mulkay. To this extent, it can be said that creationists' discourse has a similar asymmetrical accounting structure to that identified by Gilbert & Mulkay (cf. Edley, 1993; Gieryn, 1983; McKinlay & Potter, 1987; Pollner, 1974; Prelli, 1989a, 1989b; Taylor, 1996). Thus, when presenting their own claims about the world, as we have seen, they adopt the rhetoric of empiricism; and, when discussing evolutionists' views, they adopt the rhetoric of contingency, thereby attempting to undermine the credibility of evolution by linking it to various extraneous, nonempirical, personal, and subjective factors (Locke, 1994a, 1994b).

Because space is limited, I want to focus analysis on specific forms of this delegitimizing discourse. There are two aspects: the first focuses on the use of common sense as a lever of critical rhetoric, and the second on aspects of interest attribution with a specifically social dimension to them.

COMMON SENSE

Creationists often invoke common sense, something which Taylor (1996) linked to their adoption of Reid's commonsense philosophy. However mundane the appeal to common sense might be, it is—indeed, for that very reason—valuable to look more closely at the rhetorical work it does. Most notable in respect of creationists is the work it does for them in undermining evolution. An example of this comes from a pamphlet entitled "Cetacea" (i.e., whales). This is an unusually lengthy pamphlet in my sample, so, for brevity's sake, I present only two short extracts. The first opens the text; the second appears at the beginning of the conclusion. They provide the slices within which the rest of the text is sandwiched. As such, they are particularly significant to its meaning:

4.4   For a layman to attempt to review such a subject is open *prima facie* to the criticism that it is not a little impertinent when so much erudition has been lavished on this vast but specialised branch of Zoology. My defence resides wholly in a compelling wish to draw attention to the unsatisfactory nature of the presentation of this vast subject in the literature. A fault which stems it seems to me from the over-riding fact that the experts [...] wish, remorselessly, to force the discussion irresponsibly into the strait-jacket of Darwinian thought. (Pamphlet 114, p. 1)

4.5   It would be presumptious for me to express an opinion on the particular problem of the origin of Cetacea, but not being a biologist is no reason to part company with common sense or to dispense with the critical faculty. (Pamphlet 114, p. 13; spellings as original)

As can be seen, these excerpts are organized in part around a contrast structure established at the outset (4.4) between layman and expert, further referenced in the closing (4.5) by the use of common sense as an instrument of error definition. This seems to be drawing on the general meaning of common sense as something that, although by definition common to all, is especially the province of the laity; it is this tension that is being exploited.

This is a very powerful construction that works for the creationist at a number of mutually reinforcing levels. First, it invokes a specific kind of identity as an ordinary person, as someone who is precisely not a specialist or expert. This does several things. It establishes an authorial voice that is accessible to equally nonspecialist readers (likely to constitute the bulk of potential readers) and invites them to identify with it themselves. Immediately, therefore, it proposes a divide between, not just the author, but the reader and the experts, who are immediately located as the opposition. Further, it implies a role as a detached observer looking in at the exotic community of experts and seeing, with a cool eye, what they have to say for themselves. As such, it claims to be offering a more expansive view; an external eye is one that is able to offer a complete picture, one that can avoid getting bogged down in unnecessary detail and, therefore, better judge the true state of affairs. By a similar token, it is also one that is unlikely to be under the sway of particular influences from within the observed community and, therefore, able to give an undistorted view of things. Thus, the claim is established in the text that the reader is led through admittedly specialist material, but by someone whom they can trust to give them the required information without adding in unnecessary complications, someone who speaks to them in their terms—implicitly, those of common sense.

Second, the construction is preformulated (Potter, Wetherell, & Chitty, 1991) to deflect the potential criticism that defines the dilemma of the outsider. Although, an outside observer can claim the advantage of an encompassing and uncommitted perspective, they can also be regarded as ignorant and uninformed about the community's affairs (a charge, as we have seen, often levelled at creationists). Any judgment they might make can then be dismissed as uninformed, or "impertinent." The construction here, then, is designed to deflect this potential criticism. In presenting such an upfront display, the author attempts to invalidate the criticism before it can be made. In claiming to acknowledge the validity of the charge, he thereby effectively claims exemption from it. This is also implicit in his "defence". The defense itself involves a judgement of the community (that it is unsatisfactory). Thus, the implication is that, if the author recognizes the potential validity of the criticism of him as an outsider, but goes ahead in passing his judgment anyway, then he must have very good reason for thinking his judgment valid. Acknowledging the potential criticism of the judgment, then, actually helps to reinforce its validity claim, making it appear all the more justified.

And what is the judgment? Here, we see the resources of the contingent repertoire being used to construct a characterization of the behavior of the experts, which is given a further, persuasive twist through the contrast with common sense. The experts are presented as behaving contingently, by forcing things irresponsibly into a straitjacket of evolutionary thought—fitting the facts to the theory, we might say. We are left in no doubt that this is unacceptable behavior, both by its description as irresponsible (echoing the earlier unsatisfactory) and by the use of the verb force, which, matched with the description of Darwinism as a straitjacket, conveys the sense that things do not fit easily. This is reaffirmed in extract 4.5 by the talk of "dispensing with the critical faculty."

Crucial to the whole construction, however, is the reference to common sense.[5] This needs to be seen in combination with the layman versus expert contrast, such that, in effect, a parallel construction is advanced that opposes common sense, the province of the layman, to the thinking of experts. Significantly, the contrast works in favor of common sense. The implication is that the layman is actually a better judge than the expert, because she or he is unconfined by restrictions imposed within the expert community and operates immediately and directly with an instrument which, although shared by all, is lost to the expert precisely because of their specialist position. The reference to common sense, then, claims access to a more fundamental—and more critical—mode of thought than the expert's own, because the expert, in becoming an expert, has lost touch with the normality of the everyday world.

It is interesting to contrast this rhetoric with the orthodox sociological view of the consequences of expert specialization, as seen in earlier chapters. In the common sociological account, the divorce of the expert from everyday life is seen as having negative consequences for the lifeworld, leaving it "impoverished" (Habermas, 1987b) and prone to all kinds of bizarre behavior as a consequence. In the creationist's rhetoric, however, it is not the lifeworld that loses out from the growth of specialization, but the expert community itself, which has lost touch with common sense and the critical faculty that this affords.

---

[5]It is amusing—and revealing—to note that, according to Gieryn et al. (1985), Darrow's defense of evolution in the Scopes trial rested in part on the claim that evolution was "just plain common sense"!

What this shows, as far as the current discussion goes, is that the lifeworld is not without resources of its own, and nor does it lack the rhetoric of critique that intellectuals often seem to claim for themselves. Rather, this rhetoric seems to be a shared resource that may be used by the nonexpert precisely as a means of undermining the credibility of the expert and, arguably, can do so more powerfully, through appeal to the amorphous ubiquity of the powers of common sense.

Further ironies of creationist discourse arise from a consideration of their account of evolutionists' interests, to be considered next.

FORMS OF INTEREST

This discussion takes a somewhat broader form than the foregoing analysis, as I am more concerned simply to illustrate the kinds of formulation of interest that creationists' advance. My concern here is twofold: First, I wish to show that creationists have available a number of ways of characterizing the interests of evolutionists and the forces and mechanisms behind them, which show interesting parallels to the kinds of characterization of interests found in sociological analyses; second, I want to bring out the reflexive irony that characterizes much creationist discourse, especially in their employment of the resources provided by their religious beliefs as additional armaments in the struggle against evolution.

Social Forces

As I have already suggested, many of the ways in which creationists characterize the actions and beliefs of evolutionists in order to delegitimize them make reference to individual subjective factors, such as prejudice, bias, preformed attitudes, and so on, as well as stupidities and irrationalities of various kinds. Much of this repertoire focuses on evolutionists' purported selectivity in dealing with the evidence, their deceptions (pointing especially at instances of fraud), and their supposedly unwarranted readiness to accept evolution theory. From the present point of view, the most interesting thing about these types of characterization is that they direct attention at what might be called the human factors in scientific work. It is precisely this, of course, that makes them contingent. The significant point is that creationists are able to draw on such a repertoire of terms, just as much as (if not more so than) scientists themselves. They have, therefore, a handy set of critical needles with which to puncture evolutionists' claims, which can always be questioned, because they can always be characterized in such human, particularist terms.

The range of contingencies, however, does not stop there. In particular, there are a number of supraindividual, or social factors that creationists employ as delegitimations. These are of most interest here, because of the comparisons they invite with forms of sociological argument. Broadly speaking, they come in two forms: references to institutionalized processes of socialization and the dissemination of knowledge within society; and references to broader features of social, cultural and historical context. Together, they provide a not altogether unfamiliar account of ideological determination.

Socialization, Education, and the Mass Media   Creationist references to institutionalized socialization processes have two main kinds of focus: schooling and media propaganda. Extract 4.6 provides an illustration of the general tenor of this emphasis:

4.6.   [...T]he demand is unceasing that evolution theory be accepted as the only scientific explanation for origins, even as an established fact, while excluding creation as a mere religious concept!
    Such rigid evolutionary dogma, with the exclusion of the competing concept of special creation, results in young people being indoctrinated in a non-theistic, naturalistic, humanistic religious philosophy in the guise of science. Science is perverted, academic freedom is denied, the educational process suffers, and constitutional guarantees of religious freedom are violated.
    This unhealthy situation could be corrected by presenting students with the two competing models [...]. [....] This is the course true education should pursue rather than following the present process of brainwashing students in evolutionary philosophy. (Pamphlet 210, p. 4)

Here, then, a view of current educational practice as brainwashing is presented. More fully, an interrelated contrast is constructed, connecting science and true education (incorporating academic freedom), on the one hand, and dogma and indoctrination–brainwashing, on the other. Presenting the current situation in terms of the latter and opposed to the former provides a simple, but powerful rhetorical weapon to make the case for creationism. The extremity of the contrast enhances the sense of threat and urgency conveyed.

As might be surmised from the reference to constitutional guarantees, this extract is taken from a pamphlet authored by an American creationist.

As was seen in chapter 2, the public position of creationism in the United States is considerably more prominent than in Britain, especially in respect of the attempt to have creation theory taught in schools alongside evolution theory. It might be expected that, in such a context, critical characterizations of current educational practice are more likely to be advanced. Certainly a text such as this seems designed to engage with an audience for whom matters of educational freedom seem to carry special conviction (cf. Taylor & Condit, 1988).

Regardless, the example shows that creationists in general have available this kind of resource as a means of accounting for the success of evolution (i.e., due to indoctrination) and the exclusion of creationism. In a similar way, they also advance representations of the socialization of professional scientists, designed to account for their acceptance of evolution, despite the evidence to the contrary. One example of this was seen in extract 3.7, referring to the rarity of paradigm shifts among professional scientists, carrying implications of the pressures of professional socialization. A more explicit example concerning geologists appears in the next extract. This comes from a pamphlet outlining the findings of a creationist survey of the impact on the local geology of the eruption of Mount St. Helens in Washington, on May 18, 1980, which adopts a typical uniformitarian–catastrophism contrast. This extract follows a brief summary of the significant erosion caused by mudflows:

4.7. The small creeks which flow through the headwaters of the Toutle River today might seem, by present appearances, to have carved these canyons very slowly over a long time period, except for the fact that the erosion was observed to have occurred rapidly! Geologists should learn that, since the long-time scale they have been trained to assign to landform development would lead to obvious error on Mount St. Helens, it also may be useless or misleading elsewhere. (Pamphlet 252, p. 2)

In this instance, then, the responsibility for error is assigned to the training of geologists. This contrasts with extract 4.1 previously, in which attention was focused on the actions (or inactions) of a single scientist, Lyell, treated as an isolated individual. In the current example, on the other hand, no one individual is held responsible for the error; rather, it is the institutional background of the whole community of scientists that is blamed. This shows something of the flexibility of contingent characterizations: If no single scientist can be named as a target, it is still possible to accomplish blaming by addressing a whole community of them instead. There are advantages to such anonymous attacks in the position they provide for the critic as an apparently independent outsider, uninfluenced by the social forces they identify as the source of the problem (cf. the previous discussion of common sense).

In similar manner, the mass media are also a target of criticism, as in the following example:

4.8. The idea that the earth is billions of years old and that its geological development has been extremely slow and gradual is entrenched in popular opinion. This impression is given to the public by elementary level textbooks, newspapers, television and museum exhibits. However, this is misleading. There is overwhelming evidence that the history of our planet has been dominated by rapid and catastrophic events." (Pamphlet 305, p. 1)

In other words, the media have assisted in the spread of the evolutionist message, contributing to the impression that the evidence is more clear cut than it is. Especially culpable here, in the creationist view, is the presentation in the media of depictions of ape-men. One pamphlet reproduces a number of drawings of such ape-men taken from newspapers, museums, and scientific texts. Its purpose is to highlight that "reconstructions of whole heads and even bodies from mere pieces of skulls is rather a matter of artistic skill and imagination than of scientific method." (Pamphlet 151, p. 1) In particular, it shows drawings of ostensibly the same human ancestor taken from different sources, to highlight the wide variation in depictions.

It can therefore be seen that creationists have available these kinds of resources as means of accounting for the success of evolutionary thought, despite what is, in their view, its obvious errors. Processes of ideological socialization and propaganda may be identified as bearing responsibility. There is a certain irony in this, especially in the attribution of blame to the mass media. In chapter 1, it was shown that the media are among the factors identified by the Royal Society as being responsible for the spread of what they call *pseudo science*—a term that might well be taken to include creationism. This shows that there is a common, shared representation of the media as a significant source of misinformation, a representation that is readily available to serve a range of ostensibly different versions of the world—including, not least, those of sociologists.

Sociocultural Context.   As much as creationists refer to institutional processes of ideological determination, so also do they point to the wider ideological context in which evolution has developed. There are a number of aspects to this that appear in my sample of pamphlets. Many are concerned with characterizing the 19th-century cultural context in which

Darwinism appeared and developed. Broadly, these refer to the relative roles and positions of science and religion in the period, providing some kind of charting of the rise of the former and decline of the latter. Extract 4.9 provides an example of this. It comes from a pamphlet that summarized a speech made by Sir Ernst Chain (see also extract 3.8):

4.10. [Chain] states "The main theory which has dominated thinking in the biological field for almost a century is that of the Darwin–Wallace concepts of evolution and natural selection through the survival of the fittest. This mechanistic concept of the phenomena of life in its infinite varieties of manifestations, which purports to ascribe the origin and development of all living species [...] to the haphazard blind interplay of the forces of nature in the pursuance of one aim only, namely, that for the living systems to survive, is a typical product of the naive 19th century euphoric attitude to the potentialities of science, which spread the belief that there were no secrets of nature which could not be solved by the scientific approach given only sufficient time. There exist people, even today, who hold such views, but on the whole the scientists, and in particular the biologists of the 20th century, are less optimistic than their colleagues of the 19th century." (Pamphlet 204, p. 3)

Here, then, a broad characterization of the cultural context in which Darwinism became established is presented, as one marked by a "naive ... euphoric attitude to ... science" and a mechanistic conception of the world. It also establishes a contrast between this naive disposition to accept science, with the more skeptical outlook of the 20th-century, implying that this same skepticism should also be directed at Darwinism. Like extract 3.2, this trades (although not without some irony) on the common representation of the past as a bygone era, surpassed by more contemporary thinking. It thereby positions the skeptical viewpoint as the more competent and normal one. In effect, it is telling us that, if we do not have doubts about Darwinism, then there is something wrong with us. It also elicits the support of anonymous experts (scientists and biologists) to its cause, so adopting a contrasting representation to that seen in extracts 4.4 and 4.5. There, experts were positioned as relatively incompetent, abnormal, and uncritical, in contrast to the laity; here, however, it is the scientists who seem to be leading the skepticism about Darwinism—they, then, are now more critical. Once again, we see that contrasting representations can be used in different contexts to a similar end—that of undermining evolution.

Just as the 19th-century may be characterized as marked by a euphoria to science, so, too, it can be presented as religiously shallow, as here:

4.11. How real and intelligible were the parroted, second-hand Creeds and Col-

lects and Confessions of Faith—and hymns as well, that composed the
atmosphere which Darwin breathed and lived by? What did the common
people really believe about Creation, other than that it was accomplished
'Out of nothing' by Almighty FIAT! long long ago? Is it not most likely that
Lyell's uniformitarianism—'the present is the key to the past'—was inspired
by the over-repeated chant: 'As it was in the beginning, is now, and ever shall
be. World without end'—which is more apocryphal than scriptural? The
KJV—despite its Introductory boast, lamely followed the traditional Latin
vulgate of Jerome, ignoring the almost century-old translation of Erasmus,
from the Greek.
[....]
　　Darwin's 'Genesis' was [...] the KJV, a Version virtually 250 or even 500
years old, reflecting in English, the cosmogony and the biology of the Middle
Ages—before the days of telescope and microscope. It taught what the
pre-scientific scientists taught, what the common people accordingly believed, what Shakespeare believed, if not Francis Bacon himself. (Pamphlet
183, pp. 2–3)

There are a number of interesting things in this extract relevant to the current
discussion. Firstly, there is a characterization offered of the general situation
in Darwin's time—"the atmosphere he breathed and lived by." This effect is
achieved through what might (at the risk of redundancy) be called a rhetoric
of rhetorical questioning, in which the question is itself constructed so as to
advance the answer that is desired. As conversation analysts have demonstrated, questions are constructed to elicit a preferred response (Antaki, 1994).
In the indirect (Volosinov, 1973) context of a written text, however, the
construction can be expected to reflect the lack of any cointeractant to provide
an immediate reply. Thus, questions will be formulated to serve the purposes
of the author as a means of "open[ing] up a slot" (Antaki, 1994, p. 79) for the
answer the author intends to provide themselves. The shape of the question
will be designed to fit the shape of the answer to be presented.

This can be seen here. In the opening sentence, it is clear that we are to read
an answer to the question of how "real and intelligible" things were from the
shaping of the question itself—in particular its description of the Creeds, and
so forth as "parroted" and "second-hand." Evidently, such things are not very
real and so we have our answer already. This is given further emphasis in the
second sentence, structured around a "what ... other than" formulation that
again conveys the idea that the "common people" did not "really believe" very
much about "Creation." Overall, then, a sense is conveyed that the atmosphere
of Darwin's time was not a deeply religious one.

Added to this, there is a characterization of Lyell's thought advanced that
attempts to link it to a Christian influence. The suggestion here seems to be

that because religious belief has become weakened it is open to misuse. Thus, it is because the chant is overrepeated (and, moreover, apochryphal) that it could become the inspiration for Lyell's geological principle. There is an intriguing air of poststructuralism about this: The chant signifier has become uncoupled from its signified religiousness and attached to a new and quite different sign system (Lyell's uniformitarianism). Less grandiosely, the argument seems to be trading on a notion that things become devalued through overuse and, especially, their popularization, resulting in the loss of their real meaning and significance. Once again, this suggests an ironic parallel with a common argument about the consequences of the popularization of scientific ideas in the contemporary period (Green, 1985; Hilgartner, 1992).

Finally, there is a third cultural characterization at work here, concerning the translation of scripture used by Darwin. The suggestion is that the version of the Bible used at the time—the King James Version (KJV)—was inadequate, both because of its original source in "the Latin vulgate," and because of the time when it was itself written, "reflecting the pre-scientific scien[ce] of the Middle Ages." The implication seems to be that if Darwin had had available a more up-to-date Bible, then he would have found less in it to question scientifically. There are implications in this for the creationists' fundamentalist stance towards the Bible, which will be further considered in the next chapter. For now, it is enough to observe that the question of textual source provides the creationist with another way of raising questions about the context in which the theory of evolution arose.

Overall, then, this context is characterized here as a time of shallow faith, relying on parroted, second-hand clichés, overpopularized and divorced from their proper religious context, and derived, in any case, from questionable, out-moded sources. In such a context, it is no surprise that a threat such as Darwinism might flourish unchallenged—just as there are those who would claim about present-day society that it is no surprise if threats to science are flourishing when there is such widespread scientific illiteracy (e.g., Dunbar, 1995). The argument is the same; it is merely that the targets of defense and attack are reversed.

There is one further feature of creationists' characterization of the 19th-century that is worth noting. As much as this is presented as a period of declining religiosity and growing atheism, so also it is presented as one of growing materialism and, more specifically, of Marxism. Thus, for example, in one pamphlet, Darwinism and Marxism are described as an "unholy alliance" (Pamphlet 111, p. 6) and as "the twin concepts which in the intervening years have led to the development of so many of our present moral and social evils" (Pamphlet 111, p. 3)—that is, especially, atheism and materialism.

Marxism is also seen as a factor in the continuing presence of evolutionism in the 20th-century, as in the next two extracts, both of which are taken from a pamphlet linking Marxism and the notion of the primeval soup:

4.12. [...F]ollowing the publication of the *Origin of Species* and the development of Darwinism, there grew a demand for a naturalistic explanation for life's origin. This lead to the revival of the theory of spontaneous generation. The new version became known as 'chemical evolution.' One of its principle proponents was the Marxist biochemist Alexander I Oparin (1894–1980). [....] Oparin's theories were widely circulated by Professor J B S Haldane of Cambridge. Haldane, who for many years had been editor of the *Daily Worker*, was a communist and a militant atheist. [....] [...H]e proposed the notion that on the early Earth, large quantities of organic compounds had formed and accumulated in what he termed a"hot dilute soup". This concept later became known as the "primeval soup"." (Pamphlet 267, p. 1)

This is followed in the pamphlet with a detailed critique of experimental work attempting to create the materials of life from the chemistry of the primeval soup, arguing in largely empiricist terms that such an environment would have been unsuitable for even the most primitive life forms. The pamphlet then concludes:

4.13. [...T]he fundamental assumptions of the theory of chemical evolution have been found to be false. [....] It lacks both a sound theoretical basis and adequate experimental support. But why is it still being found in textbooks and taught in schools and colleges?
    The answer is that evolutionists need to believe in chemical evolution because they have excluded from their considerations the concept of God. [....] Materialistic explanations for life's origin have now become the foundation of evolutionist's beliefs and they turn a blind eye to the obvious design that exists in nature. It is of interest to point out that when Oparin proposed these ideas in the 1920s, atheistic Marxism had just been imposed on the Soviet intelligentsia. Oparin provided the necessary naturalistic explanation. He succeeded in convincing most of Western scientists also, conditioned as they were by Darwin's naturalistic explanation for the origin of species. Towards the end of his life, Oparin was awarded the Order of Lenin and made a Hero of the USSR in recognition of his service to the Marxist cause! (Pamphlet 267, p. 5)

There are many things going on in these two passages, but I only wish to emphasize the general argument. This is constructed around a connection established between "the theory of chemical evolution" and Marxism, which organizes both opening and concluding passages. In this way, the whole notion of a primeval soup is cast in the shadow of the ideological demands of atheistic Marxism. Meanwhile, it was open to ready acceptance by scientists in the West, as they had already been conditioned by the presence of Darwinism to accept the "naturalistic explanation" proposed by the marxist. Thus, despite a range of empirical data said to undermine the theory, it is still presented in school textbooks, because it agrees with the materialist's a priori assumptions. Here, then, clear answer is provided to the question of why an apparently empirically inadequate theory continues to be perpetuated. Arguably, also, an implicit appeal is being made to potential concerns about the infiltration of *Godless communism* in Western society and especially school textbooks. Regardless of this, however, the point to emphasize is that a view is given of evolutionist thought as being open to the influence of political ideology. Because the theory supposedly makes certain basic assumptions—that exclude the possibility of a deity or a spiritual basis to life—then its proponents are forced to look for material explanations for the occurrence of life even if this conflicts with the evidence, such as "the obvious design that exists in nature."

Again, then, social and cultural factors are called to account for the continued survival of evolutionist thinking. We also see in the above extract another factor—the exclusion of God—as playing a role. This type of combination of factors, focused on the rise of scientific materialism and the concomitant decline of religion, we have met repeatedly in this book under the label of disenchantment. This also seems to be present in the next extract:

4.14. In the rush and whirl of this scientific age with its multitudinous and marvellous inventions, we have lost the full sense of wonder and reverence, and
have generally ceased to admire God's superhuman works, and to give Him
thanks and glory. Even religious people may look upon the emerging chick
or butterfly as 'just Nature', without a further thought about the mystery of life
which neither time nor matter can explain. Indeed, the very regularity and
reliability of God's good gifts of sunshine and rain, seedtime and harvest,
tend to deaden man's sensibility to the miraculous in the universal provi-
dence of God. How few say Grace at meals; how few have family prayers
to-day. (Pamphlet 104, pp. 1–2)

This extract recalls the earlier discussion of the mysterious in nature, but gives to this a deeper contextualization against the backdrop of "this scientific age." Although other reasons for the loss of wonder are also advanced, science and technology are also held out for blame, in terms reminiscent of the rationalization hypothesis and its rhetoric of disenchantment. It is significant that this rhetoric is employed by the creationist. One inference it might lead to is to suggest that it adds some verification to the thesis: Even the deeply religious adopt the language of secularization. On the other hand, however, it might be suggested that the reason they do so is because they are drawing on a shared discursive resource, one which constructs a particular view of the past and its relationship to the present and, in so doing, allows certain kinds of arguments to be made and accounts given. What, then, might this rhetoric be doing for the creationist here?

It does three major things. First and foremost, it provides a means of accounting for the success of evolution. In this case, it does so through the suggestion that, in the modern age, our attention has become shifted away from an awareness of the superhuman and the miraculous, toward the distractions of science and technology. However, in attributing this to a feature of our behavior and outlook, it implies that, in actuality, nothing has really changed with the world itself. The wonders and miracles are still there for those with eyes to see. Thus, the doorway to God is still open.

Second, it provides a basis of critique of science. For the creationist (as for many others, including Weber [cf. Habermas, 1987b] himself), the loss of the sense of wonder at nature is itself an impoverishment of the human spirit. And science is held to be at least partly to blame for this situation—although, for the creationist, of course, it is specifically evolution that is the target of attack.

Third, it provides a critique of humanity itself. Although evolution is charged with some responsibility for the disenchanted state, in this pamphlet the view is taken that, ultimately, responsibility lies elsewhere. It is at this point that creationist accounting for evolution takes a distinctly moral turn. This can be seen in a later extract from this same pamphlet:

4.15. [...W]hat really is a paradox, a contradiction, an anomaly is that man who is a moral and responsible creature, is all too often immoral and irresponsible, disobedient and lawless. It is almost a miracle of evil that the only really intelligent creature should act so stupidly [...]. [...I]t is a wonder [...] of monstrous portent [...] a 'miracle' that the moralists, the psychiatrists, the lawmakers, the magistrates, the reformers, the prison authorities, can neither explain nor cure. [....] They merely repeat that man's technological achievements have outstripped his moral 'advance'.

But no one can disprove the emphatic Bible teaching that man has fallen from a loftier state of existence. [....]

[....]

Faith is that vital link with God which Adam snapped in his foolish desire to be equal to his Creator—and independent of Him. No longer could he leap, as it were, in the inspiring light of God's Presence, but must henceforth seek to satisfy his vain ambition by slow and laborious digging and delving into the earth, only to find at the end of all his toil and sweat, problems which, when God is denied, mock all his industry on the road to self-sufficiency. Denying superhuman Miracle, he is forced to invent Evolution instead. It is up to the enlightened to reverse this—rejecting imaginary Evolution in favour of the Creator and His [...] miracles [...]. (Pamphlet 104, pp. 5–6.)[6]

There is a great deal going on in this passage, only some of which I can discuss here. First, is the overall argument, which, as mentioned, attributes blame to the immorality of man, and, more specifically still, to the fall pursuant to Adam's "foolish desire." This, of course, is a reference to the biblical story of the Garden of Eden.

Also significant is the critique of experts advanced, concerning their inability to explain or cure this immoral character. Again, this type of criticism is commonplace within our society. Experts, perhaps especially those who deal with social problems, are presented frequently as being ineffectual. Here, though, the creationist can go one step further: Not only can this ready representation of inadequate expertise be drawn on, but an explanation can be given. It is because the real problem resides elsewhere entirely, arising from the fall from grace of the human spirit.

This is also used as a basis for further criticism of science, seen in the contrast referred to between "technological achievements" and "moral 'advance'". The 'scare quotes' (!) around 'advance' here are there to warn us from accepting this way of seeing things without question. This is made plain in the sentence immediately following, where reference to the fall is introduced. In other words, the creationist is telling us that rather than advancing, we have in fact fallen back. Also, the scare quotes work to call into doubt the whole structure of the contrast. The contrast works through establishing a level of relative opposition, between the *advance* of technology and the (*non*) *advance* of morality, the former outstripping the latter. In questioning the idea of moral advance, the creationist also calls into doubt this oppositional structure, displaying it as the experts' agenda, not the creationists' own. Thus, however much we may have achieved technologically, we cannot be said to have advanced by virtue of this—indeed, decidedly not, as the setup of the whole pamphlet has already established that the "rush and whirl" of modern invention has come at the cost of the loss of wonder. In this way, the creationist links our moral fall to our technological achievements, in effect reversing the sense of these as

---

[6]I should make it clear that I have deliberately presented this as a single passage, although it is actually spread over two pages of the original source. In effect, I have linked the first paragraph from page 5 of the source, with the last from page 6, with the point of abridgment as marked in the extract.

achievements. Thus, they symbolize our loss, not our victory.

This representation is further embedded through the metaphor of industry in the final paragraph. This is a very rich passage, with its play of connotations connecting Adam's moral fall from Eden and the imagery of gardening ("delving into the earth"; further compounded through contrast with "leaping in inspiring light"), to industry, itself signifying technology, in turn connoting invention, and this capped by the description of evolution as itself an invention. There is further play with the different senses of invention, referring both to a material object and a product of the imagination, so enabling the shift to describing evolution as an imaginary product. Also, describing evolution as an invention recalls the ironic talk of marvellous inventions in extract 4.14, thereby connecting evolution directly with the loss of wonder in modernity.

The core point to take from all this, however, is that creationists draw on the rhetoric of disenchantment, which they are able to deploy flexibly, making it work for them by adapting it into a basis of critique of evolution and a means of introducing the Christian message of moral fall. These two things go hand-in-hand for them, which is one reason why the argument from their critics that they misunderstand science by assessing it on moral grounds should itself be seen as a rhetorical strategy. What creationists are questioning is precisely the validity of the division between science and morality. For them, this is (part of) the problem. Employing it as a basis of critique, therefore, simply turns the effort at analysis into a game of opposing definitions.

This discussion of morality brings me to the final feature of creationist error accounting—reflexivity.

Reflexivity

There are two things to be covered here: Creationists' ironical representation of the errors of evolution in religious terms and the use of these errors as a verification of their biblical beliefs.

Before this, however, some clarification of what I intend the term *reflexivity* to mean is in order, as I seek to cover two somewhat distinct senses. One derives from ethnomethodology and refers to the process whereby members' definitions of situations become incorporated into the practical social order of the situations themselves (Garfinkel, 1967; Heritage, 1984). Within sociology of scientific knowledge (SSK), this has been applied also to the analytical techniques of analysts of science, such that reflexivity refers to the application of the same techniques of analysis that are applied to science to the studies of science themselves (e.g., Ashmore, 1989). In similar vein, Potter (1996) used the term to refer to the potential application of DA techniques to DA studies (such as this one).

A second sense of reflexivity has been developed within the public understanding of science, to refer to "the process of identifying, and critically examining (and thus rendering open to change), the basic preanalytic assumptions that frame knowledge-commitments" (Wynne, 1993, p. 324). Wynne uses this to distinguish perspectives which view science uncritically as the model of an open system of knowledge production and assume an ignorant closed-minded public, from those that do not assume that the public are necessarily any less reflexive in this sense than institutionalized science. Indeed, they may be more so.[7]

My claim here is that creationists are reflexive in both these senses. Creationists, as I have tried to show, deploy various resources, including those obtained from science itself, as a means of critique of (evolutionary) science. Thus, they are reflexive in this, Wynneian, sense. In addition, they are also able to deploy other resources to assist in the construction of this critique—in particular, arguments of a moral nature drawn from the Bible. Their use of these resources is often deliberately and overtly ironical, such that the means of critique of science they advance becomes a confirmation of the validity of the truth-claims made by this means itself. That is, their critique of evolution is used to confirm the truth of the Bible. Thus, they are reflexive also in the ethnomethodological or fact-constructional sense in which their definition of the situation effectively becomes the situation. This has a peculiar, and apparently paradoxical outcome: Creationists now appear to be both self-critical and not self-critical at the same time.

Irony. A common feature of creationist discourse is their frequent usage of words and phrases which ring with religious intonation, as, for

---

[7]There is one problem with Wynne's position from the present point of view. He sees the process of reflexivity as part of a growing public response to the so-called "crisis of late modernity." This shows the influence of Beck's (1992) notion of reflexive modernity, in which it is thought that major social institutions might come to adopt more open and self-critical mechanisms as an outcome of this crisis. However, this position seems to reproduce the very terms it is apparently trying to dismantle. Wynne's argument is that there is no basis for the assumption that the public are less reflexive, in his sense, than scientists; in fact, there are grounds for claiming just the opposite. Equally, however, I see no grounds for assuming this to be an especially recent development in public understandings of science—as the case of creationism can be taken to suggest (see chapter 2). In linking reflexivity to a current crisis, it presumably follows that things were different in the past, that perhaps science was once more reflexive or open than it has now become. This is a position I associate with the rationalization hypothesis. It seems to accept that the Enlightenment notion of science did once describe the social reality of modernity even if it does not now. This, it seems to me, mistakenly takes one common representation of science as descriptive of social reality. This is characteristic of much recent sociological analysis of the apparent social changes occurring in modern societies, as will be discussed in chapter 6.

example, in their use of the word *miracle* analyzed previously. To add to the layers of meaning involved, they also sometimes use this word with ironical intent, as in this extract:

> **4.16.** [...L]et us remember what is involved in the Evolution of man from ape. Let us examine one small point, a man's foot. An ape has an opposable great toe; that is, it works like a thumb. This is not only a great asset to the ape [...] it would also have been an asset to primative man. If he dropped his weapon in a fight he could have picked it up with his foot [...]. But this is not all; in order to lose this asset, the muscle which binds together the four toes had to make a jump round the fifth and bind all five together as in our feet. No one knows how or why this miracle happened; it is just *unthinkable*. (Pamphlet 11, p. 2; spellings as original)

Here, the word miracle is being employed in a somewhat different way than was seen before. In extract 4.3, it was used to describe natural processes, notably reproduction, in an empiricist context to describe a recurring, commonplace phenomenon, with the implication that miracles happen all the time and are there for all to see. Here, on the other hand, it is being used to highlight the empirical impossibility (unthinkability) of the phenomenon in question. On this occasion, then, a rather different kind of sense of the word miracle is deployed, one closer to what Chambers defines as, "hyperbolically, a marvel, a wonder," where hyperbole itself means, "a rhetorical figure which produces a vivid impression by extravagant and obvious exaggeration." This sense of overexaggeration is often used in order to cast something into doubt. Thus, far from being something that happens all the time, this miracle is something that will never happen.

Why has the creationist chosen this particular word to convey this sense of impossibility? Because of the irony. To a Christian, the word miracle is likely to carry the sorts of religious connotations indicated earlier—notably the idea of a wonder wrought by God. Placing this meaning in extract 4.16, the difference between the musculature of ape and human feet becomes something brought about by God, not by evolution. Indeed, its very impossibility makes supernatural intervention necessary to produce such a change. In other words, rather than the muscle jump occurring through slow processes of gradual evolutionary change, it occurred instantaneously through a creative act of God, who, in the creationist view, made humans separately from the creation of apes. The creationist, it would appear, then, is playing with the dual meaning of miracle: It is being used at the same time as hyperbolic exaggeration to expose the impossibility of the evolutionary view and, reflexively, to advance the creationist view as the only viable explanation. The fact that the evolutionist account is miraculous itself means that the creationist account is true, a rich irony, indeed!

It should not be thought that this is a unique example; irony of this kind is a recurring feature in my sample of pamphlets. It is especially common in the characterization of evolutionists' behavior as contingent. Particularly great play is made of the charge that evolutionists display types of behavior more commonly associated with religious conviction. Evolution is frequently described as a faith, and evolutionists accused of dogmatism. One pamphlet claims of Darwin that, "To this day, multitudes have followed him with a kind of religious zeal" (Pamphlet 58, p. 2). The use of the word *multitudes,* also carrying biblical resonances, adds to the ironic reflexivity of the characterization.

This reflexive turning of the tables is at its sharpest in the description of evolution as itself a religion. An example comes from the pamphlet, encountered in extract 3.9, talking of the biases of both atheistic and Christian frameworks. Thus, it is said, "It follows that the theory of evolution is a religion rather than a science, and the creation–evolution argument is a debate between the science of one religion and the science of another religion" (Pamphlet 266, p. 1). In the concluding section of the pamphlet, this is expanded as follows:

4.17. Imagine two fortresses. One is humanism. It is built on the bed-rock of evolution. Its banners bear such names as divorce, homosexuality, abortion and promiscuity. The other fortress is Christianity. The foundation of all our Christian doctrines, including redemption through Christ, is Creation. [....] The humanist strategy is [...] realistic. Their cannons are aimed at the foundational truth of Creation. If they can discredit Genesis, then those issues built upon it, like family life and the sanctity of life, will crumble and fall. If God did not create, no-one sets the rules. Why not divorce, abort, and so on? There is then no meaning to life. If the foundations be destroyed, what can the righteous do? It is time for the church to wake up to this truth and attack the foundation of humanistic philosophy, the pseudo-science of evolution. (Pamphlet 266, p. 4)

This description of evolution as part of a religion—sometimes called secular humanism—is characteristic of creationist argument and central to the reflexive irony of their position. The power of this irony can be seen in that it completes a turnaround such that evolution, rather than creation science, becomes the pseudoscience. Thus, the terms of the critique of creationism are adopted and completely reversed, with the implication that, if the critique is logically built, then, so too is its mirror image: As right as it is to damn creation science as merely a religion in disguise, then it is also right to damn evolution for the same reason. The difference is that creationists are able to accommodate the critique that their science is religiously grounded and thus turn this into an advantage—providing, of course, they are able to make the countercharge against evolution stick.

Why God Made Evolutionists.   There is one final stage in this process—biblical prophecy—as here:

4.18. We must not be surprised that so many people believe in evolution, because if we read our Bibles we can clearly see that these so-called changes in belief are predicted. We are told that in the last days perilous times shall come. Men will be seeking their own ways and will acknowledge no God at all. Is not this what is taking place today?

The theory of evolution is the thin end of the wedge towards atheism because if the Creation is a myth then why believe other parts of the Bible? This is a very insidious and crafty way of suggesting that the whole of the Bible is false.

St. Paul predicted that we should have to face such times as have now come upon us, when the great majority of the people in the world do not believe in the living God or in the Lord Jesus Christ. (Pamphlet 223, pp. 3–4)

This is the final irony. It is not just that evolution may be called a religion and pseudoscience, but that its very existence is itself proof of the Bible. This, then, for the creationist, is why God made evolutionists—as a test of faith and, paradoxically, as a proof of His Word given through the prophet, Paul. At this point, the reflexive reversal is absolute: Nothing can refute creationism, as the existence of doubt about the truth of the Bible is written into the biblical text itself, so confirming that which is being doubted. Doubt equals proof: a reflexivity of almost Cartesian dimension! With this, creationist discourse brings us face to face, so to speak, with the biblical text itself; and at this point, a new problem arises for them—how to account for alternative versions of the Bible.

SUMMARY

This chapter has considered the variety of ways in which creationists manage the problem of existence of evolution, constructing accounts that preserve the primary assumption of a single, stable reality, one that confirms the truth of creationism. In order to do so, they utilize a range of techniques and devices designed to undermine the credibility and legitimacy of evolutionary views, and confirming and legitimizing creationist views instead.

Like orthodox scientists, they draw on forms and techniques of empiricist discourse and a range of contingencies to this effect. In this way, the basic dilemma of science, represented in the form of contrasting discursive resources, is played out. The universalistic discourse of empiricism provides a repertoire of ways of representing reality as a realm of stable, asocial facts that refute evolution and confirm creation; yet, the localized discourse of contingency provides a repertoire of contextual particularities in the form of human errors and sociocultural processes of determination that enable the claims of evolution to be undermined.

Although there is this parallel, there is also difference. Creationists draw on a moral discourse as a means of accounting for evolution, one which is probably not utilized by orthodox scientists. Nonetheless, the moral discourse drawn from Christianity provides creationists with a powerful level of ironic reflexivity that orthodox science lacks. In consequence, creationists are able to turn the tables on evolution with considerable rhetorical dexterity. For not only does this add to their battery of techniques for delegitimizing evolution—notably with the charge that it is itself a religion—but also, they are able to utilize it as a confirmation of their own beliefs. Thus, the resources of this nonscientific discourse—Christianity—provide a set of critical tools in the argument, tools that are both significant and rhetorically powerful. As Wynne has suggested, then, it is not just science that has access to tools of critique; science has no monopoly on the ability to ferret out the background assumptions of alternative views and analytically deconstruct them. This is precisely what creationists do to evolution.

Thus, as I suggested earlier, creationists are reflexively critical of evolution, but in such a way as to confirm their own beliefs. This brings us to the final dimension of creationist argumentation concerning their treatment of the biblical text itself and the construction of compatibility between the Bible and science. I call this process discursive syncretism, and it is the subject of the next chapter.

5

Constructing "The Beginning"

The analysis in chapter 4 established that creationists are involved in a similar process of fact construction and argumentative legitimation as orthodox scientists. Like scientists (and other mundane reasoners), creationists are confronted with the problem of accounting for the existence in evolution of an alternative version of the way the world is. To resolve this, they adopt the classical asymmetrical accounting procedure defined by the conjunction of the empiricist and contingent repertoires—or, more broadly, the use of techniques designed to legitimize their version of reality and delegitimize the contrasting version(s) of evolution. Among the delegitimizing techniques creationists adopt are a range of moral arguments, which draw on the alternative field of discourse, Christianity. Using these, however, poses two further kinds of problem.

In a manner parallel to their confrontation with evolution, they must deal with the existence of different Christianities, which also claim to be founded on the Bible. Thus, as much as they must contend with other versions of the world, so, too, must they contend with other versions of the Word. Of particular concern to them are those versions of Christianity that interpret the Bible to fit the theory of evolution. Creationists reject such liberal adaptations of the biblical text. However, this in turn raises a further problem. As seen in chapter 4, creationists are concerned to appear scientifically legitimate and they draw on the accepted discursive techniques of science as part of this effort. They are also fundamentalists who proclaim a commitment to absolute biblical authority. Thus, somehow, they must satisfy both of these masters, science and the Bible, but they must do so in a way that will not appear to be doing what liberal Christians are doing, i.e., interpreting the Bible to suit science. How, then is this accomplished?

This chapter focuses on this question. There are two dimensions to be brought out. First, the task is accomplished through the use of a repertoire of interpretive techniques—a fundamentalist repertoire—which is designed to present what is in fact an interpretation as noninterpretive or preinterpretive. Second, central to the use of this repertoire is the attempt to persuade to the view that science and the Bible are entirely compatible. This involves a process whereby biblical terms are interpreted to suit modern scientific conceptions, but at the same time, these scientific conceptions are also interpreted to suit the Bible. I call this discursive syncretism, as it involves reading science and reading the Bible simultaneously in a mutually shaping interpretive process. This produces an understanding of science and the Bible as together constituting a seamless web of revealed truth; the version of the world and the version of the Word are effectively integrated to produce a whole—albeit not one that is itself free from diverse interpretations and usages.

FUNDAMENTALISM

In chapter 4, the empiricist repertoire was said to be marked by three features: an impersonal grammatical form; the primacy accorded to data; and the presentation of laboratory work as constrained by universally applicable rules. The upshot of this discursive form is a textual presentation of the world as if free from interpretation, as though the world itself is speaking directly to us in the text.

A very similar discursive form is at work in creationists' representations of the Bible. The Bible appears in their texts as though it, too, is free from interpretation and is also speaking to us directly. The effect of fundamentalism, then, is parallel to empiricism, only it refers to the meaning of writings, rather than actions (in a laboratory). In a sense, then, fundamentalism is a type of empiricism (or vice versa—cf. Nelkin, 1982). This prompts two related questions. First, is there a similar set of discursive devices at work in fundamentalist discourse as is found in empiricist discourse? Second, is it helpful to view the techniques of fundamentalism as simply a selected number of techniques from the wider set of fact construction, as was suggested about the features of the empiricist repertoire? An argument in favor of this is that the basic effect of fundamentalism is to convey a sense of what might be called textual factuality, the facts of what a text itself says. Like empiricism, it is an interpretation that denies its own interpretive status and so purports to present in front of the reader a preinterpreted reality, in the one case of the world, in the other of a book.

On the other hand, does the fact that it is a book that is being referred to make a substantial difference?

It does, I think, in one significant way. Following Pollner (1974, 1987), it is a feature of mundane reason that the world is assumed to be singular. In this sense, empiricism is a discursive embodiment of a prevailing norm, which, although regularly threatened, is not seriously doubted by most people most of the time. The same is not so clearly true of books, however. The idea that a book is open to many different readings, that it is, so to speak, many different books, is much more commonplace. This is not to say there is no debate. People argue over different versions of books, just as they do over different versions of reality, but the status of the arguments is, arguably, different. In the case of reality, the onus is on the person who claims reality is not singular or coherent to demonstrate it; in the case of a book, however, where the onus rests is much less clear. It is likely to vary with the particular book concerned. It is quite possible that a particular book may well be widely regarded as incoherent, as being several different things at the same time, without this being thought to be true of all books, or of books as a matter of course.

Such, I would suggest, is the case with the Bible. The Bible is, in modern society, widely regarded as an incoherent book. "You can prove anything with the Bible," is a commonplace expression. The consequence of this is to place the onus on the fundamentalist to demonstrate that the Bible is in fact a coherent, singular text, and not as incoherent as popularly said to be. This gives to the fundamentalist repertoire a somewhat different dimension than that found in the empiricist repertoire: The fundamentalist must demonstrate that the Bible is singular in the first instance, not merely operate under the assumption that this is the case and speak accordingly. On the other hand, threats to the assumption of singularity must be dealt with here, just as in the case of empiricist claims; accordingly we might expect to find a similar set of devices for delegitimizing alternative versions, whether of the world or of the Word.

Overall, then, there are points of similarity, or parallel, between fundamentalist discourse and empiricist discourse. Particularly notable is that, just as empiricist discourse draws on impersonal grammatical forms that disguise the presence of agency, so fundamentalist discourse adopts words and phrases that present the Bible as the source of agency. Similarly, just as empiricist discourse accords primacy to data, so fundamentalism accords primacy to the Bible, especially as an authoritative voice over human actions. However, there is also an important difference between empiricism and fundamentalism. Whereas empiricism makes reference to a realm of laboratory work in which actions are constrained by rules, fundamentalism makes reference to a realm of textual exegesis, where the rules are less constraining of interpretation. Consequently, there is a need to establish that the interpretation being offered is correct, that is, true to the intended biblical meaning. Central to this is to show that the interpretation follows a rule: that of textual coherence, achieved through textual gerrymandering. I will illustrate these features in turn.

THE TEXT AS AGENT AND AUTHORITY

A major concern of creationists is to delegitimize alternative versions of the Bible, especially those that, they claim, reinterpret it to suit the theory of evolution. For example, criticism is levelled in some of the pamphlets at a theological position called Double-Revelation Theory, as here:

> 5.1. [...] Double-Revelation Theory [...] maintains that God has given man two revelations of truth, each of which is fully authoritative in its own realm: the revelation of God in Scripture and the revelation of God in Nature. [....] Whenever there is apparent conflict between the conclusions of the scientist and the conclusions of the theologian, especially with regard to such problems as the origin of the universe, the solar system, the earth, animal life, and man; the effects of the Edenic curse; and the magnitude and effects of the Noahic Deluge, it is the theologian who must rethink his interpretation of the Scriptures, at these points, in such a way as to bring the Bible into harmony with the general consensus of scientific opinion. (Pamphlet 112, p. 1)

This position, which accords primacy to science over the Bible in respect of such issues, is criticized in this pamphlet for failing to recognize what are said to be the limitations of the scientific method on matters of cosmology and cosmogny (sic). The discourse used here is the now familiar one conjoining empiricist and contingent formulations to undermine evolution theory. Particular criticism is levelled at "leading exponents" of the Double-Revelation Theory, who have accepted "Gamow's 'Big Bang' hypothesis of an expanding universe," only to find Gamow himself "frankly admitt[ing]" limitations to the theory. From here, the pamphlet concludes with the following passage. I have quoted this at length, as there is much in it of interest to the current concern. To assist discussion, I have numbered the sentences:

> 5.2. [S1] This is not an isolated instance. [S2] Time and time again, Christians have been pressed into adopting some popular scientific theory, only to discover to their sorrow and embarrassment, that they had succeeded in "harmonising" Scripture to a scientific concept that was proved to be erro-

neous after all. [....] [S3] Man's understanding of the universe continues to change as he learns more and more of its intricate and marvellous structure; but God's Word never changes, for it is the direct product of an infinite and unchanging God.

[S4] [...I]t is also true that the Scriptures are inerrant and authoritative whenever they do speak on matters that overlap the so-called domains of the scientist and historian [...]. [S5] For example, there is a remarkable amount of clear Biblical evidence to show that Adam and Eve received their bodies by supernatural, direct creation (rather than by an evolutionary process) [...].

[....]

[S6] *Evangelical Christians must therefore strongly challenge the popular notion that modern science provides an independent and equally authoritative source of information with the Bible concerning such doctrines as the original creation*, the Edenic curse, and the Noahic Flood, and that science alone is competent to tell us *when* and *how* such things occurred (or even *whether* they occurred!), while the Bible merely informs us "in non-technical language" as to *who* brought these things about and *why*. [S7] The truth of the matter is that the Word of God not only provides us with the only reliable source of information as to the *when* and *how* of these great supernatural events (to say nothing of the *who* and *why* in each case), but also tells us why the unaided human intellect is utterly incompetent to arrive at the correct answers in such matters, (cf. Rom. 1: 18–23; 3: 11; I Cor. 1: 19–29; 2: 14; Heb. 11: 1–6; II Pet. 3: 3–5).

[S8] We are far from denying that God has given men a revelation of Himself in the material universe, for the Bible definitely teaches this in Psalm 19: 1. [S9] Furthermore God commanded Adam to "subdue" the earth (Gen. 1: 28), and we may presume that this command finds partial expression in the marvellous inventions and discoveries which God has permitted to His creatures.

[S10] *But there are many vitally important truths that are completely outside the scope of scientific investigation.* [S11] Cosmogony, cosmology, and metaphysics, in the ultimate sense of these terms (and no other sense is truly valid) are impossible apart from God's special revelation in Scripture. *[S12] The true Scientist, therefore, no less than the true theologian, must confess with David "Thy word is a lamp unto my feet, and a light unto my path ... in thy light shall we see light."* (Ps. 119: 105; 36: 9).

[S13] In view of all this, the Christian may have perfect confidence that science can make no ultimately fruitful discoveries that are not in perfect accord with the clear and obvious teachings of God's Word. (Pamphlet 112, pp. 4–6)

This passage displays all the basic elements of the fundamentalist repertoire. First, the Bible is presented as an agent in its own right. This is apparent in the general thrust of the argument that the Bible should not be harmonized with science, which already implies that it has existence and meaning independent of human interpretation. In addition, specific words and phrases are used recurrently to refer to the biblical text as an independent actor: "the Scriptures ... speak" (S4); "the Word of God ... provides ... also tells us ..." (S7); "the Bible definitely teaches" (S8). The talk of "clear Biblical evidence" (S5) has a similar effect. It is the Bible doing these things, not a human reading of it.

Breaking this down further, two things are apparent. A number of different ways of referring to the Bible are used (the Scriptures, God's Word, etc.). Particularly notable are those references that represent it as acting directly on behalf of God. The text then becomes the voice of God speaking directly to us. This is seen in such passages as: "God's Word never changes, for it is the direct product of ... God" (S3); "God's special revelation in Scripture" (S11); and "the clear and obvious teachings of God's Word" (S13). Similarly, the formulation "God commanded Adam" (S9) effectively transforms a two-step representation process into a single step, as it is an abbreviated version of a statement of the form: "It is written in the Bible that God commanded Adam." In reducing this to the direct, "God commanded Adam," all sense of representation vanishes from the description, such that the events appear before us as things that have actually taken place. Thus, all sense of interpretation, at either level, similarly disappears.

Also notable are the recurring references to the Bible as not requiring any interpretation, because it is "clear" (S5; S13); "obvious" (S13); "definite" (S8), and so on. A similar effect is achieved with the contrastive formulation in S6, "the Bible merely informs us "in nontechnical language." Several things are happening here. First, this is a view of the Bible we are being invited to reject, as can be seen from the contrast with the view that "science alone is competent" to judge on matters of origins; in effect, then, we are told, the Bible does more than merely inform. Second, this particular construction, merely inform, uses both the minimization (Potter, 1996, p. 187) mere, and also the verb inform itself to accentuate the denigration. Informing contrasts with the other terms used to describe the Bible as speaking, telling, and teaching. Although these terms may be treated as synonyms of informing, in this context it conveys a relatively weaker sense. Substituting any one of these verbs for this inform shows this: Merely teaching and merely informing are two different things. In particular, here, inform connotes a sense of information without great significance—information but not knowledge, we might say.

This sense of comparative weakness is also enhanced by the phrase "in nontechnical language," itself placed inside quotation marks. Like the scare quotes discussed in chapter 4, these quotation marks seem to be working as a distancing device, signifying less a specific quotation and more that this way of representing the biblical text is attributable to others, but not the author himself (he is male). The general direction of the argument invites speculation that the author wants to reject the implied contrast between the language of the Bible and the language of science. Technical language is language that might well be associated with science especially. Thus, to describe the Bible as nontechnical is to imply a contrast with science—and, perhaps, even to go a little further and convey a sense of inferiority, so that the nontechnical text is judged less advanced, perhaps, than the technical one. This resonates also with the reference in S2, to "some popular scientific theory" (see also S6). In these contexts, the popular seems to be derogatory, conveying the idea of being swept along by the crowd thoughtlessly. Something that is popular may also be considered nontechnical (perhaps especially if it concerns a scientific theory). Overall, then, a sense is conveyed of the Bible as an agent that does impart information, but does much more than this also, including things not necessarily of a nontechnical—that is unscientific—nature.

We have seen, then, how the Bible is constructed in this stretch of discourse as an agent. It is also presented as a primary source, that is, as an authority. This is made quite explicit in S4—"the Scriptures are inerrant and authoritative whenever they ... speak." But there is more to the constructive work involved than this, for a very good reason. It is one thing to maintain that the Scriptures are inerrant whenever they speak, but quite another to ensure that they are speaking when you want them to in the ways that you want them to. In this case, this is accomplished through constructing a reading of the Bible around two mutually reinforcing contrast structures, so as to close down successively the interpretive options potentially available.

The first contrast is established in S3, between "man's understanding of the universe," which "continues to change," and "God's Word," which "never changes." There is a potential pitfall here, however. In modern culture, being unchanging is often represented as a negative trait, implying inflexibility, datedness, unprogressiveness, even ignorance—and these are all common attributions made of fundamentalists.[1] Thus, the construction here is designed to resist these associations and, moreover, to reverse them, so that it is changeability that becomes the weakness. This is achieved

---

[1]In the sociological literature, fundamentalism is usually characterized as a reaction against exogenous change, in particular as an essentialist reaffirmation of established group belief (Davie, 1995). The term itself originates in an early 20th-century American Protestant movement, seen as attempting to resist an encroaching modernism (Bruce, 1988; Kepel, 1994).

through a second level contrast between what the unchanging Bible can do that changing science cannot: Specifically, it can explain science in a way that science cannot explain the Bible. This comes about through a four-step circle of argument.

Thus, first, in S7, there is a reference to what the Bible tells us about the incompetence of the "unaided human intellect." This recalls, it is interesting to note, the type of contrast employed by Gilkey concerning the logical limits of science. Once again, we see the creationist employing similar rhetoric to the social scientist. In this context, however, this is used to assert the authority of the Bible ("the only reliable source"), as an entitlement to "evangelical Christians" (S6) to challenge double-revelationists.

Second, in S8, is a reference to further biblical teaching that "God has given man a revelation of Himself in the material universe." The significance of this rests in the fact that the reverse is much less probable—that is, although we might be happy to accept the idea that a book might provide such a teaching (whether or not we agree to accept it), we are likely to be less prepared to accept that the material universe teaches us to expect to find a revelation of God in a particular book. The book, then, has the advantage to this degree and, thus, greater authority. Also, this presentation offers a particular meaning of the biblical text. Reference is given to Psalm 19:1, which is, in the King James Version, "The heavens declare the glory of God; and the firmament sheweth his handywork." This is interpreted by the creationist to mean that God has revealed Himself in the material universe. Thus, the Bible is interpreted even as the rhetoric of fundamentalism denies any interpretive presence.

Third, this takes on a more specific focus in S9. The text invites us to interpret God's command to Adam "to *subdue* the earth," as including "inventions and discoveries," that is, modern science and technology. This is a common reading of this particular biblical passage, but its familiarity should not be allowed to disguise its status as an interpretation. It is a very specific meaning read into a very broad textual reference. That the creationist is aware of this seems apparent from the change in modalization in the clause, "we may presume that." With this, the authorial voice is weakened and modified from one masked by the fundamentalist voice of biblical agency, to one that invites the reader to participate (we) in constructing the meaning of the text. In this way the fiction of the Bible speaking for itself may be maintained, despite the knowledge that interpretation is at work. In inviting our participation in the work of construction, it also invites us to agree to maintain a suspension of disbelief in the

Bible as an autonomous, independently existing voice, in order to enable the argumentative narrative to continue.

The final step in this narrative is the reconnection to the limits of science, in S10 through S12. Here the contrast between the Bible and science becomes fully explicit: There are things that the Bible can do which science cannot. The Bible can speak on matters of "cosmogony, cosmology, and metaphysics," things that are impossible without it. Although this restates the first step, completing a circle of argument, it does so in the light of the constructions in steps 2 and 3, which advance a specific interpretation of the Bible that constitutes the defense of the circle. In particular, it is an interpretation that links science and technology directly to God's command, thereby making them an outcome of God's will and intention and dependent upon Him for their existence. Further, because God's will itself is expressed directly in the Bible, the Bible becomes confirmed as the greater authority over science—a view that is capped by delineating, in S12, the behavior and attitude of the "true Scientist."

In this way, then, the potential accusations that may follow from advocating a source of knowledge that is unchanging are deflected. The Bible is presented as an absolute, but one that does not by virtue of this make further scientific work pointless. Rather, as S13 tells us, "discoveries" may still be made, but they will only be "fruitful" if they are in "perfect accord" with the Bible. The Bible, then, does not stop science—just the opposite, it makes it possible. It might even be that God has commanded us to do science, in the command to Adam to subdue the earth. This will only be successful if it is true science—and we will know it is true, because, if so, it will confirm the Bible!

The primary point, then, is that this kind of connection between science and the Bible requires close interpretive work of the biblical text. However, this is presented in the terms of the fundamentalist repertoire, so that it does not appear to be an interpretation—or, at least, its status as an interpretation is hidden or denied. In this way, the interpretation is legitimized, and other interpretations are undermined.

This brings up a further issue, however. As much as fundamentalist discourse may parallel empiricist discourse in its style of legitimation, so also does this leave open the question of accounting for the existence of alternative versions. If it is the case, as the fundamentalist repertoire so insistently has it, that the Bible speaks clearly and obviously and that, as is stated in other pamphlets, its meaning is plain, why then do so many Christians appear to get it wrong?

CHRISTIAN ERRORS

The contingencies creationists use to account for alternative versions of the Bible are broadly similar to those used to account for the existence of evolution. This is unsurprising when it is recognized that these techniques are designed to undermine the legitimacy of the alternatives and, as such, make use of characterizations that are features of commonsense discourse in general. Thus, a similar type of asymmetrical accounting procedure is at work in creationists' theology as in their science; that is, just as they present their account of the world in empiricist terms drawing on contingencies to represent the evolutionist alternative, so, too, do they present their account of the Word in fundamentalist terms, drawing on contingencies to represent Christian alternatives. As before, the range of contingencies used is wide, and there is no space here to give examples of them all. Rather, I concentrate on accentuating the parallels with the empiricist–contingency structure. Thus, I again look at the use of common sense as a rhetorical device, at references to social factors as means of accounting for error, and at the significance of reflexivity. As it happens, each of these features is displayed particularly clearly in one particular pamphlet from my sample. Accordingly, I provide examples from this one source, although I emphasize that each of these features, plus a range of other contingencies, is to be found in other pamphlets as well.

Common Sense. Earlier, it was seen that common sense was mobilized as a means of discounting the views of evolutionist experts. Similarly, it is also used as a means of undermining nonfundamentalist accounts of the Bible, as in extract 5.3. Here, the author is criticizing what he (he is male) refers to as the "nonliteral interpretation" of the Bible, also called the "Literary Framework Hypothesis":

5.3. Our final objection to the Non-Literal Theory is that it is far too complicated. Every teacher knows that you begin with the simple and move on to the complex. This principle can be clearly seen in the Bible, too. Prose in the historical books leads on to poetry in the Psalms, philosophy in Ecclesiastes, prophecy in Isaiah, and finally the difficult 'visions' of Ezekiel and Daniel. But the non-literal school would have us believe that right at the beginning of His revelation God has placed a conundrum as hard to solve as any in the whole Bible. Listen to this comment on Genesis 1: '*The writer has given us a masterly elaboration of a fitting, restrained anthropomorphic vision, in order to convey a whole complex of deeply-meditated ideas*' (Henri Blocher, 'IN THE BEGINNING', 1984).
Anyone who has tried to teach the elements of Christianity to uneducated

people will recognize the utter impossibility of explaining to them why God's first words should be that gobbledegook, rather than plain statements of fact easily intelligible in every language to all nations—as the pioneer missionaries believed. The Literary Framework Hypothesis is a house of cards carefully constructed by academics in the airless atmosphere and artificial light of a theological library. We need to open the windows and allow a good strong blast of common sense to blow it down. (Pamphlet 260, p. 5)

This passage presents an explicit contrast between the constructions of academics and common sense, given added color by the metaphor of airlessness. The metaphor is well chosen, too, as it draws on commonplace representations of academics as dry and dusty pedants, so deeply buried in their books as to have forgotten what real life is all about.[2] Thus, the views of academics are presented as overintellectualized, "far too complicated" for their own, or anybody else's, good. Also notable is the further contrast between academic gobbledegook and the Bible's own "plain statements of fact easily intelligible in every language to all nations." The fundamentalism here is presented in a particularly extreme form, that has something of the quality of what was earlier called extreme redundancy (chapter 4). Describing plain statements as easily intelligible has a certain circularity, which in conjunction with the two extreme case formulations (every language and all nations), doubles the repetitive quality. This emphasis is further enhanced by the choice of description. Gobbledegook is American slang for official jargon and, although it may commonly be applied more broadly than this, makes for a particularly pertinent application in this instance. In effect, it positions the nonliteralist view as the official argot of an in-group of professional academics (notably theologians), in contrast positioning fundamentalism as the language of everybody else—including, of course, the creationist. Thus, nonliteralism is undermined as an exclusive, elitist ideology.

Finally, the reference to common sense further works this contrast in a similar manner seen previously, exploiting the sense of academics as a distant body of specialists, divorced from everyday reality and living, literally, in a fantasy world of their own making (a house of cards). Further, a contrast is also made between academics and "anyone who has tried to teach," the anyone an extreme case formulation that, by default, includes all who are not academics. Once more, therefore, the sense is conveyed of academic theologians as a caste of distant mandarins, irrelevant to everyone except themselves.

---

[2]This is a charge sometimes levelled at academics by other academics, especially, perhaps, from those who express opposition to relativism and related positions, as for example, in the treatment of Derrida at Oxford (discussed in Forrester, 1996). See also Dunbar's (1995) comments on the humanities.

To reemphasize the earlier point, then, this shows that common sense is both a resource and itself an instrument of critical argumentation. Crucially, this resource includes characterizations of academics, experts, and so on, designed to undermine their authority and the credibility of their view-points. Creationists use these resources to position themselves on the side of everyman, thereby seeking to enhance the legitimacy of their views by emphasizing their commonality (cf. Pomerantz, 1986).

Social Factors.   However, as much as it may be contrary to common sense to take a nonliteral reading of the Bible, it would seem from the next extract that this has been a recurring practice throughout the ages:

5.3.   [...R]einterpretation of Scripture is an old game. **"Full well you reject the commandment of God that you may keep your own tradition ... you hold the tradition of men ... making the Word of God of none effect through your tradition which you have handed down ..."** (Mark 7). We find this attitude of Christ to the Old Testament uniformly consistent [...]. He never re-interpreted Scripture. He simply quoted the words as being in themselves perspicuous, intelligible, and meaningful, in the plain sense of common speech. Why did this offend the Pharisees? They were certainly fundamen-talists. They believed in an inerrant Book. But they had re-interpreted the words to suit their own life-style.

As we move on through the New Testament we find again and again a similar resistance to new truth, or rather, to old truths rediscovered. *'O fools and slow of heart to believe all that the prophets have spoken!'* [....] Once again, Bible-believers were blind to Bible truth because of the current philosophy—in this case expectation of a conquering Messiah.

We can follow the same theme through church history. [....] [...T]he schoolmen [...] re-interpreted St Paul's words to fit in with the medieval ecclesiastical system. It was 'all a matter of interpretation'. So it was in the days of John Wesley. Anglican prelates disapproved of open-air preaching, in spite of obvious precedents in the Acts of the Apostles. Baptist elders [...] re-interpreted Christ's command to suit the *laissez-faire* philosophy of 18th century England. When George Muller and Hudson Taylor affirmed that it was possible for Christian work to be supported *'by prayer alone to God alone'*, Christian businessmen laughed them to scorn. The promises had always been there, in Matthew 6, but *'little faith'* had reinterpreted them as being contrary to experience.

[...U]ncomfortable doctrines are jettisoned to prevent them rocking the boat. Outward profession of conformity to Scripture is retained even when practice and teaching differ widely from its pattern. And not infrequently there is heavy reliance on tradition [....] [...T]he Fourth Commandment was [...] carved on stone—*'Six days shalt thou labour ... for in six days the Lord made heaven and earth ...'*—but 20th century disciples have 'read it many ways'. (Pamphlet 260, p. 1)

Here, then, we have a view of the Bible as having been repeatedly reinterpreted throughout its history; even, in the case of the Old Testament, by the Pharisees, before the coming of Christ. Particularly notable is the way in which the objection to this practice is entitled, through reference to Jesus' own criticism of these same Pharisees, in the quotation from Mark 7. It is also from these words of Jesus that the cause of the error is identified, in rejecting "the commandment of God" and keeping to "the tradition of men." This is taken as generally indicative of Christ's attitude to Scripture, the extreme case formulation, "He never reinterpreted Scripture," acting to suggest this was His normal practice and, hence, normalizing the fundamentalist's claim. Further, Christ is constructed as Himself a fundamentalist, "simply quot[ing] the words as being in themselves perspicuous, intelligible, and meaningful, in the plain sense of common speech." The use here of common speech positions Christ—and also the creationist who is adopting His voice—as an everyman, much as we saw the term common sense doing before. This is accentuated further by the contrast with the Pharisees who, although also fundamentalists, had "reinterpreted the words."

The precedent set by Christ is used to authorize the identification of further examples of such reinterpretation and their explanation in terms of the traditions of men. This in itself is massively interpretive; it extracts a general meaning from Christ's words, which is then made to apply to a range of specific instances throughout Church history. It is worth setting this out as a list, in order to display the range of meanings drawn from this one phrase. Thus, Christ's words are taken to mean: "reinterpretation" in general; being "blind to Bible truth because of the current philosophy"; fitting in "with the medieval ecclesiastical system"; disapproving of "open-air preaching"; suiting "the laissez-faire philosophy in 18th century England"; reinterpreting promises "as being contrary to experience"; jettisoning "uncomfortable doctrines ... to prevent them rocking the boat"; retaining "outward profession of conformity to Scripture," but differing from it widely in "practice and teaching"; frequently relying heavily on tradition; and reading the Fourth Commandment "in many ways" (a reference to Kipling).

The point of this is not to suggest that these different characterizations are inconsistent or incompatible, but only that a great deal has been extracted from one biblical reference (itself having a localized meaning in its original context), and that this requires interpretive work. In this case, the interpretive work is such as to suggest that numerous meanings drawn from the Bible over the course of history have been motivated by factors external to the text itself and, specifically, factors that reflect aspects of prevailing ideologies. This itself is taken to apply at a range of levels of action and belief, from the relatively local and specific—such as disapproving of open-air preaching—to the relatively wide and general—such as fitting the medieval ecclesiastical system, or 18th-century laissez-faire philosophy. In each case, however, this is presented as a reinterpretation of Scripture, in contrast to its "perspicuous" meaning, associated with plain common sense.

Thus, a similar kind of asymmetry to that seen before is at work in the discourse. One reading of the Bible is presented in fundamentalist terms and others in contingent terms, just as one reading of reality is presented in empiricist terms and others as due to various contingencies. In addition, the same kind of contingency—ideological interest—may be used to account for both erroneous readings of reality and of the Bible.

Reflexivity. The previous passage also shows a certain ironic reflexivity in employing the biblical text itself as a basis for rejecting readings of the biblical text. In chapter 4, I gave as a further example of reflexivity the use of quotations from the Bible to explain the existence of evolution. This same reflexivity is employed also to account for the presence of nonliteral readings. Thus, following the critique seen before (extract 5.2) and a brief discussion of Darwinism, reference is made to the same biblical citation:

5.4. One of the "signs of the times" is [...] the TIMES ATLAS OF THE BIBLE (1987) which does not show Mount Ararat on any map! The religious publishing world has decided to expunge every trace of that uncomfortable story which thunders God's wrath against sin. The Apostle Peter predicted just such a time: *'In the last days scoffers shall come .... they deliberately overlook the fact that the world that then was, being over-flowed with water, perished.'* (II Peter 3). (Pamphlet 260, p. 5)

It is perhaps debatable that, here, these scoffers are intended to include also the nonliteralists, but, given that this is the concern of the bulk of the pamphlet, a case for it can be made. Thus, there is a dual critique here: On one level, it is an example of accounting for the error of nonliteralists by reference to the role of the mass media as purveyors of propaganda; and, on another, as a reflexive accounting of this practice itself by reference to biblical prophecy. As was seen before, then, the Bible provides creationists with the means of accounting for its own rejection, and the rejection becomes confirmation of its truth.

Which Bible?

This returns us to the fundamentalist repertoire. Earlier, I said that fundamentalists have to contend with the problem of guaranteeing the internal coherence of their textual authority. The existence of different interpretations of the biblical text in itself raises this issue to the fore, just as it does in the case of competing interpretations of the world. The rhetorical resources of contingency provide one means of managing this problem. It is also dealt with in another way, however, through the construction of versions of the Bible that demonstrate its coherence. However, in the case of the Bible, this presupposes the resolution of a preliminary problem concerning which version of the Bible is taken as constituting the Bible in the first instance.

The problem here is that different versions of the Bible exist in two senses: Not only are their different interpretations of any one Bible, but there are actually different Bibles, in the sense of different canons, the combination of books used to make up the Old and New Testaments. The question of the canon is less a problem in itself, as creationists will simply follow the Protestant canon, established with the King James Version (KJV) of 1611; however, this is complicated by the existence of different translations into English. Thus, there are already many different Bibles, even before the question of whether any given Bible can be said to be internally coherent is broached. How do creationists deal with this?

The simple answer, based on my sample of pamphlets, is that they do not. Although there are numerous quotations from and citations to the Bible in the pamphlets, it is rare to find any mention of the specific version used. From checking a sample of these quotations, it would appear that the text used is the KJV. However, it will be recalled from extract 4.11, that this version of the Bible was criticized as being out of date "reflecting in English, the cosmogony and the biology of the Middle Ages [...]. It taught what the pre-scientific scientists taught [...]." (Pamphlet 183, p. 3) Since 1611, a number of other English translations have been produced since, notably the Revised Version (RV), appearing between 1881 (NT) and 1885 (OT); the American Standard Version of 1901, revised as the Revised Standard Version (RSV) between 1946 (NT) and 1952 (OT); translations by Moffat in 1913 (NT) and 1924 (OT); and the New English Bible (NEB) versions of the NT in 1961, OT in 1969, published complete with Apocrypha (i.e., noncanonical books) in 1970.

However, although these various translations consist of the same collection of books, the same cannot be said of one other reference to a specific version of the Bible in my pamphlets. This is the LXX, also known as the Septuagint, a version of the Hebrew Old Testament translated into Greek some 200 years before Christ (traditionally by 72 scholars—hence Septuagint). Certain books of the Septuagint were later repudiated by Jews and followed in this by the early Protestant reformers. These, along with some other books about Christ, also considered inauthentic by Protestants, now constitute the Apocrypha. It is interesting to consider this particular reference, which comes once again during the critique of nonliteralism and concerns the date given for the creation (said by the creationist to be "some 6,000 years ago"):

> 5.5. Flavius Josephus, a Jew of the first century AD [...] had unequalled opportunities of investigating and understanding the culture and traditions of his own people. How does he handle the early chapters of Genesis?
>   1. *'Moses says that in just six days the world and all that is therein was made... Moreover Moses, after the seventh day was over, begins to talk philosophically....'* In other words Josephus is saying that chapter 2 may be a bit mysterious, but in chapter 1 there is no hint of any mystery at all. He obviously takes the days as literal.
>   2. *'The sacred books contain the history of 5000 years...'* This is conclusive proof that the Jews of Josephus' day added up the figures in Genesis 5 and 11 to make a chronology. He later states: *'...this flood began 2656 years from the first man, Adam.'* (Both computations are based on the LXX text). (Pamphlet 260, pp. 2–3)

Of main interest here is the reference to Josephus' computations being "based on the LXX." Josephus' reading is defended as literal and this used as a defense for literalism now. As is apparent, though, Josephus' text was not the same as any contemporary accepted Protestant text, because of the exclusion of the Apocryphal books and because Josephus would probably have read the early Greek translations, which have since been translated first into Latin and then into English, to obtain the KJV, itself since revised into the RV, RSV, NEB, and so forth. This does not necessarily mean that the computation of the genealogical history of the Bible would differ, regardless of which translation or version is used—although, it is notable that Josephus, writing in the first century A.D., gives a figure of 5,000 years, when the creationist, writing some 2,000 years later, says "by all the laws of language it is certain that Genesis tells of a six day creation some 6,000 years ago" (Pamphlet 260, p. 2), or, about 1,000 years out from Josephus. Regardless, the point here is that the rhetoric of fundamentalism allows considerable flexibility in the treatment of texts as equivalent or not. Treating texts as equivalent or not is a way of justifying a claim, of providing, in the cases we have seen, authorization of a claim concerning the age of the earth, and a basis of rejection of Darwinism. Which Bible to use, then, and/or how to represent the authoritative status of the text, varies with argumentative intent. The nature of the Bible varies just as does the nature of science, as seen in chapter 3.

Having said that, it is also a feature of creationist discourse that they often found their interpretations of biblical passages (especially from Genesis) on direct translations from the original Hebrew.[3] Some examples of this will be seen shortly. First, however, there is the construction of the internal coherence of the Bible to be considered. Previously, I argued that an extra task confronts the fundamentalist in having to establish that the Bible, as their primary source of reference, is coherent, in addition to establishing that their reading of it is also the correct one. These two things—coherence and correctness—are, to an extent, mutually validating. It can be argued that a correct reading is one that demonstrates the text to be coherent; but, demonstrating that a text is coherent can also be taken to mean that the reading must therefore be correct. Thus, coherence is itself a persuasion to correctness. Something of this can be seen in extract 5.6, from a pamphlet entitled, "Genealogies and early man":

5.6. The early chapters of the Bible, with Adam, Eve and the Serpent in the Garden, men living for nearly one thousand years, and Noah and his floating zoo, are widely discounted as history. They have been replaced by alleged ancestors of modern man going back 3.75 million years in Africa. Yet Biblical genealogies link Adam and Noah with characters whose historicity is beyond dispute. This paper studies the genealogies, considers their implications and discusses some of the difficulties encountered.

   There are two family trees of Jesus Christ. One in Luke's Gospel chapter 3 traces the ancestry back through David and Abraham to Adam 'the son of God.' The other in Matthew chapter 1 gives the descent from Abraham through David. Luke's line from Abraham back to Adam agrees with the genealogies found in Genesis chapters 5 and 11, and would have been copied from that source. Matthew's and Luke's line between Abraham and David agree together and with genealogies found in I Chronicles chapters 1 and 2 [...]. However, from David to Jesus Christ, the two gospels differ completely. Matthew's is the royal line of descent whereby Christ inherits the throne of His 'father' David, and the term 'begat' signifies heritage as well as sonship. For example, Salathiel begat Zorobabel, but I Chronicles chapter 3 tells us that Zerubbabel was the nephew of Salathiel. Under Jewish law, as our own, a nephew would inherit where a man died without issue. Matthew's Messiah inherits the throne of David through 'Joseph, the husband of Mary,

[3]In this respect, their procedure is not altogether different from my own, in that they do make efforts to supply the reader with as much of the evidence on which their interpretations are based as they can. The reader is invited to draw from this their own conclusions about the validity of the procedure.

of whom was born Jesus, who is called Christ'. 'Was born' is the feminine singular, contending that Joseph had no part in this birth. A comparison of the names reveals that, though Matthew almost certainly took his material from I Chronicles, he deliberately left out five generations in order to give patterns of fourteen steps. Since Luke's line is much longer than Matthew's between David and Christ, it is probable that Matthew omitted more generations still. This is permissible since 'begat' implies a line of descent, succession or birthright rather than a father to son relationship.

Luke is concerned with the humanity of Jesus and so traces the Lord's family tree from his mother Mary back through David's son Nathan and then right back to Adam. Zorababel and Salathiel are names which feature in this list too, but do not refer to the princes of those names in Matthew's line. Other common names occur more than once in Luke's genealogy, such as Matthat, Joseph and Levi, as one would expect. (Pamphlet 219, pp. 1–2)

Here, then, an argument is advanced to establish the consistency of the various genealogies that appear in the Bible. The setup for this concern in the opening paragraph is notable for its linking of the consistency of the Bible with the consistency of reality. The argument seems to be that, if it can be shown that the pre-Flood genealogies (Adam to Noah) link with the later ones covering "characters whose historicity is beyond dispute," then a case for treating these early characters as also genuine historical figures will have been made, thereby casting doubt on evolution. Thus, a singular Word and a singular world go hand in hand; if the coherence of the Bible can be shown, this in itself bolsters the case for its historical accuracy.

However, coherence alone is not enough. The Bible also has to be shown to justify the faith in Christ as the Messiah. Thus, the genealogies must be matched in such a way that they provide proof of Jesus' claim to be "the son of God." As this is a position attached particularly to Adam in the text, what needs to be shown is that Jesus can also claim the title, by virtue of descent from Adam. Thus, the problem is also one of demonstrating proof of Christ's significance in the history recorded in the Bible. The history and the religion must be made into one.

This is achieved by three crucial interpretations of the biblical text. First, there is the meaning attached to the word begat as something that "signifies heritage as well as sonship." This does two major things in resolving the problem of coherence. It provides a way in which the apparent discrepancy between the separate genealogies in Matthew and Luke may be overcome; and a way in which the genealogy in I Chronicles, 3, may be matched with these others. The dual meaning given to begat allows Matthew's genealogy to be taken as one of inheritance, or birthright rather than birth, as such; while, Luke's can now appear as the genealogy of Jesus' birth (his humanity). Similarly, it also enables Matthew's descent from Abraham to be connected with Luke's, which takes the line all the way back to Adam, thereby linking together the OT lines with those of the NT. Thus, Jesus' position as the proper inheritor from David and Abraham is established, which also positions Him as a figure whose "historicity is beyond dispute," by association with figures like David and Solomon. Further, in also connecting Jesus and these other figures with the line from Adam to Noah, these characters also take on the same quality of indisputable history, and Jesus Himself becomes the inheritor of Adam, equated with him as "the son of God."

The meaning of begat performs a second function to further tie this bundle of connections together. The precedent used to defend the dual interpretation is the case of "Salathiel and Zorobabel." These names, taken from Matthew (although they appear also in Luke, a point I will return to shortly) are identified with the characters in I Chronicles, "Salathiel" and "Zerubbabel." That is, Zorobabel is identified with Zerubbabel, enabling the connection between Matthew 1:12 and I Chronicles 3:17–19 to be made. Clearly enough, then, the creationist argument rests on this identification, although there is nothing given to defend the claim that they are one and the same. To further complicate the matter, Luke also lists Salathiel and Zorobabel in his genealogy of Jesus. He, however, asserts that "Zorobabel ... was the son of Salathiel" (Luke 3:27, KJV) whereas the creationist identifies Zorobabel with the Zerubbabel of I Chronicles and states that "Zerubbabel was the nephew of Salathiel." This potential inconsistency is resolved by claiming that the names in Luke's list "do not refer to the princes of those names in Matthew's line," but are rather other people who just happened to have the same names, which are common. In other words, Luke's Zorobabel is a different person to Matthew's, whose Zorobabel is really Zerubbabel of I Chronicles—all of which goes to show that consistency is constructed through interpretive procedures applied to the text, procedures which, at this point, rest on ways of deciding whether or not things are identical! This can be thought of as a type of process of categorization and, as such, as Billig (1985) has argued, carries rhetorical weight.

The third issue here concerns the manner in which Luke's line is interpreted in contrast to Matthew's. Matthew's is taken to be the line of heritage, rather than blood, and it is Matthew who uses the word begat. Further, Matthew uses the formulation "was born," which, we are told, is feminine singular in tense (presumably referring to the original Greek), thus "contending that Joseph had no part in this birth." In contrast, Luke's genealogy is taken to trace "the Lord's family tree from his mother Mary back through David's son Nathan," even though Luke mentions Joseph and not Mary. Be that as it may, what the creationist is doing here is attempting to resolve these apparently discrepant lines of descent by making use of the fact that Jesus had two parents (Joseph and Mary), but also, in a (for Christians, highly significant) sense, only one human parent (Mary). The distinction between heritage and bloodline can then be equated with the distinction between the two parents, a move that becomes especially significant in Jesus' case because of his unique birth. Thus, it can be claimed that each line of descent refers only to one or other parent: Matthew's to Joseph, justified by the word begat and his shorter genealogy connecting Jesus by inheritance to the Kings of Israel; and Luke's to Mary, justified by his lengthier genealogy, connecting Jesus by blood to Adam.

Overall, then, it can be seen that the coherence claimed for the Bible is a construction arrived at via a number of interpretive steps, involving the meaning attached to the word *begat,* the selective connection (and disconnection) of certain names, and the moves enabled by Jesus' multiple parentage. By this complex of interpretation, the Bible can be claimed to be internally unified and, from this, to be a genuine historical document, recording actual past events. This brings me to the final issue of concern: The interpretive accomplishment of compatibility between the Bible and reality, as given by science.

The Scientific Bible

In addition to managing the problem of competing versions of the Bible, creationists must also demonstrate that the Bible is compatible with science—or, rather, with creationist science. To do so, they engage in complex and subtle interpretive work, using the full range of their discursive resources, to dovetail the meanings drawn from both biblical and scientific sources. These constructions pervade their writings, and the best I can do here is to provide a flavor of the interpretive work involved through a few select examples. My main intention at present is illustrative. Discussion of the implications of this feature of creationist discourse must wait until chapter 7.

It will come as no great surprise that a good deal of the interpretive work creationists engage in to establish the compatibility of the Bible with science is devoted to the matter of origins, especially the events of the Genesis creation week. I begin, then, with one example taken from a pamphlet entitled, "The first seven days of the universe":

5.7. *In the beginning God created the heaven and the earth.* The Second Law of Thermodynamics tells us that there was a beginning, since if the universe were infinitely old it would now be completely run down, dark, cold and dead. The First Law tells us that the universe is incapable of creating itself, so a Creator must have been involved. The evolutionary scenario of the spontaneous eruption of everything out of nothing, and increasing order out of that initial explosion, is a philosophical proposition rather than science. (Pamphlet 299, p. 1)

The first sentence here is a direct quotation of the first sentence of Genesis 1:1. The remainder of the passage, then, is an interpretive elaboration drawn from this opening, single sentence. It is interpreted in such a way so as to be conformable with the First and Second Laws of Thermodynamics, two of the foundations of contemporary physics.

The First Law is the law of conservation of energy. It is a law of conservation in processes of physical exchange, such that everything that was known to be present at the start of a physical process in a closed system must be accounted for in the end product resulting from that process. It is also held to be reversible. Applying this to the universe as a whole, therefore, it follows that all the energy of the universe that is currently in existence—which from a materialist standpoint constitutes everything—must always have existed, albeit not necessarily in its current physical state. This raises the question of where the matter and energy came from in the first place. As can be seen, the creationist interprets the First Law to mean that "the universe is incapable of creating itself" and this in turn is taken to imply that "a Creator must have been involved" (a view disputed by some contemporary physicists, such as Hawking, 1988).

The Second Law of Thermodynamics is concerned with entropy. Entropy can be thought of as a measure of relative disorder, which, thermodynamically speaking, effectively refers to the energetic level of a system, relative to its local environment. The more ordered a system, the higher its level of relative energy. Over time, the Second Law predicts that the entropy in the system will increase, that its relative energetic level will decrease, and the level of disorder will increase. For all practical purposes, this can be thought of as an increase in the level of heat in the general surroundings, which, eventually, will leave an overall even temperature. Applying this to the universe as a whole, the inference may be drawn that, given enough time, the universe will eventually decline into a uniform state of relative disorder at an even, fairly low, temperature. If this should happen, all processes of energy exchange would cease, as there would be no local system with a higher energy level relative to its surroundings to provide the necessary dynamic. Thus, as the creationist puts it, the universe will be "completely run down, dark, cold and dead."[4]

This is then used as an argument against the view that the universe is infinitely old and in support of the view that it must have had a beginning. In effect, therefore, the opening sentence of the Bible is taken to be a statement that contains both the First and Second Laws of Thermodynamics—the one, because it is taken to imply that a Creator was needed, the other because it is taken to mean that there must have been a beginning of the universe. The biblical text, then, is interpreted to appear to be in agreement with the foundations of contemporary science, these foundations themselves being interpreted to suit the biblical text (most notably in the idea that God must exist).

Also of note is the contrast established with "the evolutionary scenario," presented as involving "the spontaneous eruption of everything out of nothing and increasing order." This is a version of the common creationist criticism of evolution, that it is incompatible with the Second Law. In consequence, evolution is labelled a philosophy, rather than science, on the grounds that it conflicts with both the First and Second Laws. Thus, not only is it the case that the Bible and science are mutually confirming of each other, but also that they are mutually in agreement in being incompatible with evolution. Implicit, then, is the fundamentalist stance previously seen that to accept the Bible is to reject evolution.

The first sentence of the Bible is also the concern of the next extract. This is especially interesting for two reasons: first, because it provides a clear example of the construction of a literal reading; second, because this reading is arrived at via a return to what is called "the original Hebrew."

5.8. Those who think that science has disproved much of the Bible ridicule the idea that the Book has anything important to say about how things began. Nevertheless, if Genesis or any other part of Scripture is to be criticised as scientific, let it be judged on what it actually says; let it be given a fair hearing!
Turning to the first words of the Bible we read: *"In the beginning, God created the heaven and the earth"* (Genesis 1:1). The original Hebrew has the plural, *"heavens"*; we are not to think that there is only one place called Heaven. [....]
The Bible reveals [...] of Creation [...] the Person and the purpose behind the manner and means. Creation is ascribed to the Word, Who was God, in the beginning; to the same Person Who is called both the Son of God and

---

[4]Hawking (1988: 145) called this the "thermodynamic arrow of time." The creationist view of entropy is criticized in Patterson (1983). Trefil & Hazen (1995) saw this view of "heat–death" as very much a 19th-century one, long displaced by other theories.

Lord Who *in the beginning founded the earth and made the heavens* (John 1:1–3; Colossians 1:13–17; Hebrews 1:10).

Some writers use the expression, *creation out of nothing*, but this is virtually meaningless and is not in accordance with the basic sense of the word translated "*create.*" Creation was *out of God*: the whole universe of matter, energy, space and time itself, being made perceptible out of His own unlimited energy.

Natural processes—the way things behave—are seen to obey two fundamental laws. **Nothing has ever been observed which contradicts these laws**, which state firstly that **matter and energy can neither be created nor destroyed**; and secondly that **the energy in any closed system is always being "locked out" in such a way that it becomes no longer available for constructive and self-ordering processes.**

Creation, being the absolute origination of everything apart from the Creator, was unique. It cannot have been effected by the physical processes now seen in operation; Creation did not obey the two laws just described. The God of the Bible is eternal, outside time and space, and transcendent. The act of Creation was supernatural; not conflicting with the laws of science […] but transcending them. Science, by its own definitions can never know God, for He is beyond time and space) nor comprehend Creation. (Pamphlet 245, pp. 1–2)

This stretch of discourse shows (among other things) very clearly the manner in which fundamentalism is employed as a legitimizing technique designed to present detailed and extensive interpretive work as preinterpretive. The opening paragraph does the basic groundwork here in asserting that the Bible should be judged on what it actually says, conveying the idea that there exists a meaning of the biblical text accessible independent of a reading process. Similarly, the injunction to let the Bible "be given a fair hearing" uses the commonplace discourse of the courtroom to suggest that judgments have been made of it on the basis of some kind of misrepresentation, which is correctible by letting the Bible speak for itself, just as we might want to insist that an accused person be allowed to present their own case. Thus, in likening the Bible to an accused, it is invested with agency and implied that what is to follow in the text is the Bible's own testimony, that we are about to be witness to the preinterpreted, readerless reading of the text. The reading that does follow, then, is authorized as the Bible's own voice speaking.

What does follow, however, is an extensive elaboration of the book's meaning. After quoting the opening sentence in English, we are immediately told that this sentence is, in effect, a mistranslation of the Hebrew, which uses a plural, heavens. This is brought out in order to propose a connection with a later citation from the NT, which uses the plural, rather than the singular. In this way, the claim is justified that the Bible tells us of the purpose of creation, as the connection forged is to particular passages that refer to God's primary creation in a context concerned with Jesus and His ministry. The connection is further emphasized by the sentence linking God, the Word,[5] the Son of God, and the Lord, all also linked to the beginning. In other words, God determined from the start that Jesus would come and his salvation project is thereby determined to be the purpose of creation. Similarly, Jesus' coming is made to be implicit in the Bible's opening sentence. Or so a fair hearing shows.

Further interpretive work follows this, addressing the error of those who talk of creation out of nothing. The reasons given for this error are interesting. First, it is said to be virtually meaningless; second, it does not accord with the "basic sense of the word translated create." The full sense of this, however, only becomes clear in the following paragraph. Here, we meet another statement of the First and Second Laws of Thermodynamics, with the appended extreme case empiricist claim that "nothing has ever been observed which contradicts" them (another common creationist emphasis; Hawking [1988, p. 103], however, said otherwise). Thus, to talk of creation out of nothing is to contradict these laws, whereas, creation out of God does not, because God has "His own unlimited energy" with which to perform such creative work. So, claiming that the basic sense of the word *create* implies God, rather than nothing enables this link with the laws of physics to be made. In effect, the statement "out of nothing" is said to be meaningless, because it is made to contradict the First Law; whereas, the statement "out of God" is neither meaningless nor contradictory, because God supplies the energy. The meaning to be extracted from the Bible, then, is made to fit in with science.

Equally, though, the meaning of science is given by the Bible, as can be seen in the final paragraph. Here, the act of creation is interpreted as transcending "the physical processes now seen in operation." In consequence it need not obey the Laws of Thermodynamics, but neither is it comprehensible by science. This, then, deflects the potential question of how something that is beyond science can be compatible with it, rather than in contradiction. Both science and God claim to be universalistic; how, then, can they be married, without imposing limitations on the validity of either? The creationist accomplishes this through a crucial interpretation of the meaning of *create*. Because creation is "out of God," God is positioned both

---

[5]A reference to John 1:1, "In the beginning was the Word, and the Word was with God and the Word was God" (KJV).

"beyond time and space," and so outside of the realm of science, but also as a part of time and space, which are "made perceptible out of His own unlimited energy." Once again, therefore, important interpretive work is needed, in this readerless reading of the Bible, to make it such that what it actually says is compatible with science.

One further example from this same pamphlet helps to show even more clearly how the interpretation of particular words through a series of steps, is worked to construct a sense of compatibility between science and the Bible. I discuss two extracts in conjunction:

5.9. In Genesis 1:2 we read: *"And the earth was without form and void"*; **tohu** and **bohu** in Hebrew. The same two Hebrew words are rendered *confusion* and *emptiness* in a context of judgement, namely Isaiah 34:11. [....] From the correspondence of thought in Isaiah 45:18–19 and from the other occurrences of the words under consideration, it appears that, at this time, the Earth was dark, uninhabitable and empty.

   Some scholars believe that the words quoted from Genesis 1:2 should read, *"And the earth **became** without form and void."* This is because the Hebrew word used here may mean *become*, but the experts are still not agreed on this matter. Some believe that a great catastrophe took place in between the event recorded in Genesis 1:1 and the movement of the Spirit of God (verse 2). It is thought that this catastrophe followed the judgement and fall of Lucifer (Satan) and his angels. There are certain passages in Scripture which could refer to this event, but we cannot be absolutely sure about them. It seems certain that the Earth was covered by water, *"the deep"* at this time. Many geological features such as mountain ranges and the oldest sedimentary rocks (the Pre-Cambrian) may have been formed then. (Pamphlet 245, p. 2)

A little further on, it is said:

5.10. "The creation of light (verse 3) is followed by a series of divisions. Light is divided from darkness; the waters by the firmament; and the dry land from the seas. The word *"firmament"* is better rendered "expanse"—something stretched out; it was possibly a part or parts of the many-layered atmosphere, above which exists a water vapour layer *("the waters above the firmament")*. The presence of such a layer could have accounted for the warm climate which is thought to have existed all over the world at one time. There is also geological evidence for the existence of all the land in one place at an early stage in the Earth's history." (Pamphlet 245, p. 3)

The construction of compatibility here can be seen first in 5.9 through the contrasting manner in which the meaning derived from tohu and bohu and that concerning the verb shift from was to became are dealt with. The verb shift is left open as, we are told, "the experts are still not agreed" as to the correct translation. The question here is why this matter is raised at all? For, in so doing, a potential risk is introduced that the notion of a readerless reading of what the Bible actually says may be called into question. If we rely on experts to make final judgements about the meaning of the text, then it becomes immediately apparent that the text is not open to a pure, preinterpretive, literal reading. Moreover, this positions the reader as non-expert and, therefore, as unable to make proper sense of the Bible directly, without expert assistance.

More curious still, this note of doubt about the meaning of the text contrasts with the relative certainty with which other meanings are asserted. In particular, tohu and bohu are interpreted—on the basis of something called a "correspondence of thought" between parts of the Bible—to mean that "the Earth was dark, uninhabited and empty." This claim is modalized (Latour, 1987; Potter, 1996) by the expression "it appears that," thereby introducing a note of doubt, but arguably, only a minor one. Added to which is the strongly modalized claim that "it seems certain that the Earth was covered by water," based upon the single noun "the deep." Thus, although doubt sufficient to trouble experts is expressed about the translation of one verb form, there is little hesitation in assigning other meanings to words, to present a picture of the earth as a dark, wet wilderness. Moreover, this is then presented as grounds for forging a connection with modern scientific concepts, regarding the "geological features," notably the "sedimentary rocks" of the "Pre-Cambrian."

Why, then, the inclusion of expert disagreement? What seems likely here is that the creationist is attempting to manage a problem of marrying the biblical text with traditional Christian teaching. The story of Lucifer being cast by God from Heaven does not appear in Genesis, although there is mention of such an event in the New Testament, Book of Revelation (Rev. 12:7–9), which talks of a "war in heaven." The timing of this in relation to the six days of creation does not seem to be made clear, but the event is central to Christian soteriology, because it provides a means of accounting for the existence of evil (including evolution) in a world created by a God of absolute goodness. Thus, there is something of a dilemma for fundamentalists: On the one hand, they must base their theology only on what the Bible says; on the other hand, they want what the Bible says to conform with certain accepted Christian beliefs, which includes Lucifer's challenge to God. Thus, in this instance, opening up doubt about the literal meaning of the text allows the potential for this piece of Christian theology to be incorporated into the Genesis story, even though it is not directly mentioned.

In maintaining a distance from the interpretation, however, the creationist avoids the risk of appearing to read too much into the text and in so doing contravening the fundamentalist injunction of literalism.

Having raised the issue, the creationist also cues the reader that this interpretation is a possibility. This in itself may be enough to elicit agreement to the interpretation, especially as the possible objections to it are presented only in vague terms of expert disagreement—and given its centrality to Christian soteriology. Vagueness can be used to support a claim as it makes it difficult to challenge (Potter, 1996). Further, in showing knowledgeability about this lack of agreement among scholars, the author positions himself as scholarly, thereby providing entitlement for the interpretation, even as he distances himself from it. His lack of commitment to it itself helps to validate it as a possible reading.

In turn, this further cements the legitimacy of the reading given to those words about which no scholarly disagreement is registered. If disagreement is mentioned in one case, but not in others, it helps to enhance the sense that these readings are beyond dispute and simply state what the text actually says. Disagreement over matters of the source of evil may be allowed, but there can be no disagreement over the compatibility of the Bible with science.

This emerges in extract 5.10, where the meaning of the word firmament is outlined. First, we are told unequivocally that this is "better rendered "expanse."" which opens up certain possibilities of meaning, pushing the sense in one direction rather than others, a process continued in next defining expanse as "something stretched out." Again, certain meanings are opened up, and others are closed down. Finally, it is suggested that this could be referring to "a part or part of the many-layered atmosphere," which includes "a water vapour layer." Thus, the firmament is interpreted to fit in with the scientific understanding of the atmosphere as layered and containing a substantial proportion of water. And the biblical reference to "the waters above the firmament" is also brought in line with a claim drawn from modern science that the earth once had a uniformly warm and humid climate.

One additional extract, also concerned with the firmament further shows up this interpretive work and also gives some idea of how creationists attempt to account for the biblical flood. This comes from a pamphlet which discusses each of the opening verses of Genesis in turn. Beginning at verse 6:

5.11. 6. And God said, Let there be a firmament in the midst of the waters, and let it divide the waters from the waters.
    7. And God made the firmament, and divided the waters which were

under the firmament from the waters which were above the firmament: and it was so.

8. And God called the firmament Heaven. And the evening and the morning were the second day. The word firmament denotes an expanse and is used in this chapter in v.15 to mean outer space and in v.20 to mean the atmosphere. God here makes the atmosphere and places some of the water above it. When the first sputniks were sent up in the late 1950s it was found that above our atmosphere is a deep volume of space at a temperature of many hundreds of degrees, the thermosphere. It was pointed out that this space, now empty, was capable of holding a vast amount of water vapour. Hydrogen bonding allows water vapour to vary in density and to be lighter than air at very high temperatures. It would seem that God protected the early earth from harmful radiation by this water vapour canopy. The increased atmospheric pressure would lower some metabolic rates (breathing, heart beat, etc.) and would mean that wounds would heal more rapidly. These factors may have contributed to pre-Flood longevity, so that Adam lived for 930 years. This would also explain the source of the 40 days and nights of rain during the Flood. Water vapour absorbs infra-red rays so this canopy would give a greenhouse effect. The earth would be almost uniformly warm and moist, ideal for growing the giant vegetation as found in the fossil record. With no arid heat nor polar ice, there would be little wind. Wind is first mentioned in Scripture just after the Flood. This lack of wind would result in an absence of rain, as reported in Gen 2:5,6. [....] Although the Bible does not specify a water vapour canopy above the atmosphere, it is hard to imagine any other explanation for the waters above the firmament." (Pamphlet 299, pp. 2–3)

In this passage, the interpretive reasoning process that forges links between the Bible and science is clearly set out. From a few sentences in Genesis, an elaborate interpretation of the nature of the antediluvian earth is developed. This turns particularly on the meaning assigned to the specific word, firmament. Especially notable is that this word is openly declared to have two different meanings. It is said to denote an expanse, the indefinite article allowing the flexibility to translate the word as either outer space or the atmosphere, depending on what is considered best to suit the context. This extra flexibility allows the biblical text to be interpreted to conform even more closely to modern scientific understandings.

The conjunction of science and religion is elaborated further here in the understanding of the atmosphere, drawing on the discoveries of modern satellites and contemporary ideas about the behaviour of water vapor. This is used to develop a relatively full account (compared to the density of Genesis) of how God originally created the atmosphere, with a protective water canopy, and what the implications of this would have been for life on earth. These are constructed in a manner suited to other information drawn from the Bible, treated literally in matters of the longevity of the ancients (such as Adam's "930 years"), the Garden of Eden environment (implied in the reference to the "uniformly warm and moist" planet and the "giant vegetation"), and the lack of wind and rain. Equally, however, this description of the earth's environment is designed to conform to creationists' reading of science. Thus, the reference to "giant vegetation as found in the fossil record" provides an empiricist justification for the atmospheric theory advanced. Similarly, other empiricist claims are made about the effects of such an atmosphere on "metabolic rates" and the healing of wounds.

Finally, the other feature incorporated is Noah's Flood. The proposed water vapor canopy provides a means of explaining "the source of the 40 days and nights of rain during the Flood." Once again, therefore, the biblical text is given an interpretation that enables it to fit with an understanding drawn from science, and the science itself is also selected and interpreted to suit the biblical story. There is a combined process of textual and ontological gerrymandering to dovetail the religion and the science neatly together. The Bible is read in a specific way, so that certain possible meanings are read into it, and certain others read *out*; similarly, science is read in a comparable way—for example in respect of the effects of such a water canopy on human (and animal) metabolic rates. Possible negative effects of living under increased atmospheric pressure are not considered, only the positive features to accentuate the paradisiacal state prior to the Fall. Thus, certain meanings are accentuated, and others are ignored, in such a way as to construct the required compatibility.

In this way, a syncretism is achieved, through critical, rhetorical, and discursive processes of selection, interpretation, representation, and argumentation. This process has implications for the issues raised in chapter 1 about the public understanding of science, discussed again in chapter 7. First, however, some further discussion of the discourse of rationalization is needed.

6

Science—Variation in Species

With chapter 5, the empirical analysis of creationism is completed. In this chapter, attention returns to the issues of a more theoretical nature raised in chapter 1. Accordingly, the focus of analysis shifts away from the writings of creationists to those of sociological theorists of modernity, and the issue of concern is not the rhetorics of science used by creation scientists, but those used by these sociologists—at least, insofar as these have a bearing on the relationship between science and the wider public. Thus, this chapter builds on the basic argument presented in chapter 1 concerning the rhetoric of rationalization widely employed within the corpus of texts that constitute contemporary sociology.

The basic aim is to explore the workings of a particular kind of representation of modernity, as an essentially scientific society. I considered two examples, in the writings of Gellner and Merton, in chapter 1, as a backdrop to the discussion of the explanations of creationism in chapter 2. In this chapter, the analysis is expanded to include three other theorists whose work has been, in one way or another, influential in contemporary sociology, and provides useful illustration of the working out of the dilemmatic of science through the ambiguities and irresolution of the discourse of rationalization.

The central argument runs as follows. Over the past 200 years, modernity has been consistently characterized, whether potentially or in actuality, as a scientific society. However, there is within this characterization a (not so) hidden dilemma. For, on the one hand, science is presented as something that stands in unique relationship with the form of a particular society—modernity; but, on the other hand, a much broader claim is made for the validity of science as a universal form of knowledge, applicable regardless of the particularities of local social and cultural conditions. Thus, science is both the knowledge of one particular society and the knowledge of all societies (i.e., no particular society). This dilemma has worked through especially in sociological theories of modernity, in which the role and position of science is assigned a particular significance on a number of levels, theoretical, methodological, and empirical. Indeed, it could even be suggested that this dilemma has become the dilemma of contemporary sociology, as it struggles to find a way of upholding the validity of its own knowledge claims (qua universalist science), against the seemingly irresistible pull of the abyss of relativism (Giddens, 1993), inspired by the reflexive application of the sociology of knowledge.

One proposed resolution of this dilemma seems to have been especially significant in the development of 20th century sociology: The rationalization hypothesis. However, this thesis does not successfully resolve the dilemma, but only defers it, shifting it onto other sites, the traces of which can be discerned in the writings of contemporary sociological theorists. In particular, it emerges in the puzzle over the apparent resurgence of traditional beliefs (e.g., Kepel, 1994) and the apparent appeal of so-called antiscience movements (Holton, 1992, 1993), or countercultural and New Age alternatives to the scientific–technological society (Bloom, 1991; Heelas, 1996). These seem to embrace a mixed variety of superstitions, alternative beliefs, and various other irrationalities, things that were supposed to have been eclipsed and eradicated by the gradual spreading of scientific enlightenment and its torch-bearer, disenchantment.

The attempt to account for this puzzle places severe stress on the thesis of rationalization, exposing the dilemma within. This can be seen in two related areas in particular: in the characterizations of science, which seem to shift and move in unresolved ways; and in the characterizations of the relation between science and the wider cultural order, which appear to be defined in an increasingly arbitrary fashion.

I endeavor to demonstrate this through an analysis of some of the writings of three contemporary social theorists: Habermas, Bell, and Lyotard. I take them to be representative of three main lines of development of the rationalization hypothesis in its most general form in current sociological thought, in critical theory, postindustrial society theory, and postmodernism. As before, I treat these authors as representative of a discourse. Their writings manifest the shifts and stresses referred to previously. Despite their various differences in theoretical traditions, philosophies, and politics, they are nonetheless engaged in a common struggle to make the rationalization hypothesis fit the puzzling facts of modernity—bearing in mind that these facts only emerge as puzzling, because of the characterizations of science and modernity that the discourse of rationalization advances. I treat their writings, then, as symptomatic of the problematics of this discourse. Consequently, what follows makes no pretense to an exhaustive discussion; it deals in each case only with selected works, chosen for the extent to which they aid my diagnosis and that are thus analyzed with the intent of clarification. It is less an attempt to engage with their arguments and more an attempt to dismantle certain aspects of the discourse with which they construct their arguments. This will nonetheless lead me to adopt a stance toward their analyses of the current condition of modernity—that they are inadequate, precisely because they are constructed from the resources afforded by rationalization.

I address all three—Habermas, Bell, and Lyotard—in turn. In each case, I present a reading of certain of their writings, which aims to identify what they have to say about science and the relation between science and the wider public. I argue that these theorizations are systematically ambiguous in both of these respects, although the ambiguities are more obvious in some cases than others. Part of the argument is that the ambiguities become more apparent as we move through the list of authors, as the struggle to contain the underlying dilemma of science becomes more difficult to the extent the theorist seeks to accommodate the growing tendency to represent modern culture as highly diverse and fragmented. From this, I claim that the position of science as a cultural resource provides a means of resolving the problems identified and a better way forward for the sociological study of the role and position of science in modernity.

Habermas

In chapter 1, I argued that the attempt to account for public criticism of science in terms of rationalization leads to the introduction of arbitrary divisions into the representation of science and culture. Thus, Gellner (1974) advanced a distinction between serious and playful strands of culture, while Merton (1968a, 1968b) sought to maintain an essential unity of culture as a value system, but forged a distinction within science between its utilitarian applications and its fundamental nature as an embodiment of the ideal critical democracy.

The construction of such distinctions is a response to the underlying dilemma of science in modernity. This dilemma arises at one level as a question over where exactly to locate science within the spheres, strands, subsystems, or whatever, of society. The place that science occupies has to have very special properties: It has to be a place that is both common to all societies and yet also in some way unique to modernity, in order that science can be both universal and local. Moreover, it has to be a place that is both material and ideological, as science is held to be both material, in the form of technology (especially military and industrial production), and ideological–cultural, in the form of a system of knowledge and belief about the world.

Thus, the discourse of rationalization postulates implicitly the placement of science in a social location that has the dual property of being both of the social and yet beyond the social. Unsurprisingly, perhaps, this gives rise to difficulties in particular theorizations attempting to join the social and asocial dimensions of science. Gellner's strategy of splitting culture locates science primarily materially, isolating it from a section of the ideological–cultural domain as ironically pathological. Merton's strategy of splitting science, on the other hand, divorces the technological, utilitarian applications of science from its essential nature located within a unified cultural value system.

Habermas' strategy was to split both science and culture. To this extent, his position can be thought of as a synthesis of Gellner and Merton. In consequence, however, the ambiguities that beset rationalization arise at both levels in his writings.

To deal first with science. Like Merton, Habermas (1971, 1987a) distinguished between science as a technical, or instrumental, form of knowledge and science as a practical (i.e., moral–normative), or communicative, form. For Habermas, however, this distinction is of a more fundamental nature. Whereas for Merton, it is used simply to try to preserve the essence of science from certain kinds of social usage, for Habermas it is itself essential: There are inherently two different forms of science.

Forging this distinction, however, raises an ambiguity over the relationship between the two types of science, as can be seen in Habermas' earlier writings. Here, the two types of science, termed technical–instrumental and moral–practical, were associated with the "empirico-analytical" (Habermas, 1987a), or natural sciences, and the social, cultural, and historical sciences, respectively. This distinction allowed Habermas to generate a deeper critique of the social position of science than Merton, as he is able to propose the dominance of the one form of science over the other, particularly in the hegemony of positivism in the social sciences (Habermas, 1987a). At the same time, this dominance is reflected or embodied at a broader social—specifically economic and political—level, in the isolation of the capitalist economic subsystem from noninstrumental considerations, and the increasing displacement of democratic decision making by technically qualified experts (Habermas, 1971).

However, the distinction between technical and communicative forms of knowledge is ambiguous. Communicative forms of knowledge are those that derive from and are concerned with symbolic–interactional processes, which for Habermas (1987a) have as their prototypical instance, language. Technical forms of knowledge, on the other hand, are understood as arising from the direct dealings of the human "behavioral system" (Habermas, 1971, 1987a) with the empirical world of nature; natural science, in other words, is a direct extension of technology. Technology itself develops through processes of trial and error experimentation directed ultimately toward the achievement of optimum efficiency of economic production. Thus, natural science is understood in hypothetico-deductionist, or falsificationist terms (Habermas, 1987a).

The problem with this, however, is that it is not clear how natural science can be insulated from communicative science in this way. In presenting natural science in essentially falsificationist terms, Habermas implicitly denied the validity of the Husserlian critique of positivism, which rests upon the claim that even natural science is dependent on social, and therefore symbolic, interactions. Natural science is also social and linguistic, not merely behavioral and technical. Yet, Habermas did accept the Husserlian critique of positivism and based his own critique of "the scientistic self-understanding of scientism" (Habermas, 1987a) in part upon it (cf. Bernstein, 1976).

This difficulty arises from the attempt to resolve the dilemma of science within the confines of rationalization. Within rationalization, the rationality of science, regardless of type, rests ultimately upon empirical grounds: Science tells us the truth of reality, whether natural or social. Its procedures are ultimately grounded on the principle that we can have direct access to reality; it is this that makes it rational. But this is an implicit denial of the social basis of science. Rationalization attempts to accommodate this social basis by proposing modernity to be a society in which social action itself has become increasingly rationally (i.e., purposively) oriented and thus increasingly in conformity with the procedures of science. However, this seems to entail the inevitability of the reduction of the social to the technical, something which Habermas wishes to avoid. To rescue the social, he splits it off from the technical, but in so doing threatens to undermine the resolution of the fundamental dilemma of science that rationalization has advanced.

Habermas apparently recognized some difficulty with his prior formulation of the relation between the two types of science, as this has undergone some reformulation in his more recent writings (Habermas, 1984, 1987b).

In this new version, natural science and technology are no longer altogether isolated from the communicative; rather, they are drawn in as one level of argumentation within the social lifeworld. The basic problem, however, remains.

In *The theory of communicative action* (1984, 1987b), Habermas presented a modified understanding of the social as, fundamentally, effectively coterminous with the lifeworld, which itself constitutes the seat of communicative rationality. All things, therefore, in basis stem from the lifeworld. The course of human social development is understood as a gradual unfolding of the structural potentials contained incipiently within the lifeworld. These structural potentials themselves refer to the ability to raise and contest validity claims, which ability, in essence, defines the rational (Habermas, 1984).

Habermas (1984) identified a number of types of validity claim, which embody different levels of human communicative action and that find their practical realization in the form of different spheres, or subsystems of society. The conception of validity claims can be seen as an expanded and modified form of the earlier distinction between the technical and the practical. Of particular importance here is a division Habermas introduces into the technical itself, between propositional statements and teleological (i.e., goal-directed, or purposive) action. Propositional statements are statements which make assertions, or claims, about the nature of the world and are therefore open to be tested for their truth content. Teleological actions are actions oriented to success in the sense of changing the world in accordance with one's aims and intentions and can be judged according to "criteria of ... efficacy" (Habermas, 1984, p. 87).

In essence, this amounts to the introduction of a division between scientific knowledge and technology such that scientific knowledge can now be thought of as a form of action oriented to understanding, and technology is "technically useful knowledge ... converted ... into techniques of production" (Habermas, 1987b, p. 168), that is, action oriented to success. Nonetheless, it remains the case that, in his view, scientific knowledge is still identified with technology; they are said to be "the *same* knowledge [used] in *different* ways" (Habermas, 1984, p. 11; emphasis original). At some point in human social development, however, they become uncoupled:

> the cultural tradition must interpret the lifeworld in such a way that action oriented to success can be freed from the imperatives of an understanding that is to be communicatively renewed over and over again and can be at least partially uncoupled from action oriented to reaching understanding.

This makes possible a societal institutionalization of purposive-rational action for generalized goals .... (Habermas, 1984, p. 72)

This distinction, thus, seems to have the effect of locating science outside of the subsystems of production and control, placing it instead firmly in the cultural system—that is, what remains of the lifeworld, after these other subsystems have become uncoupled. Science—natural science—thereby becomes comprehensible now as assessable by the same processes of argumentation as apply to other cultural spheres of the lifeworld. Thus, instead of dividing natural science from social science in the same sense as his earlier work, in the revised version, natural science, too, is understood as a product of communicative—that is argumentative—processes in much the same way as other forms of communicative action (which include, especially, the moral–practical and aesthetic spheres of culture).

This seems to amount to a restatement of Merton. Unsurprisingly, then, it suffers from the same basic problem of irresolution between the empirical and the conventional. It is one thing to claim that at some point the argument must stop and become partially uncoupled; precisely the problem, however, is to identify when and how this happens. This question has been central to the research work of sociology of scientific knowledge (SSK); as yet, it remains unclear that such uncoupling can be said ever to occur (Collins & Pinch, 1993; Latour, 1987; Potter, 1996).[1]

The basic dilemma, then, remains. It emerges also at the level of the public understanding of science. Once again, it is notable that there has been some modification in Habermas' position on this issue. In his early statements, he appeared to be quite unequivocal in seeing positivism as the dominant understanding of science within both the scientific community and society at large. This is particularly so, because he saw the dominance of positivism as providing the means of legitimation of the interventions of the welfare state in the capitalist market, but in a way that remains insulated from meaningful democratic involvement. This is achieved by the presentation of political decision making as a purely technical matter, thereby circumventing the need for wider public debate. Thus,

... propaganda can refer to the role of technology and science in order to explain and legitimate why in modern societies the process of democratic decision-making about practical problems loses its functions and "must" be

---

[1]Further, there is some indication that Habermas (1984) understood action oriented to success in essentially objectivist terms, stating that it is "susceptible of being judged by a third person in respect to fit and misfit" (p. 87) with the world. It is this, however, that is the question (i.e., can a third person, a nonparticipant, properly judge the success of a participant's actions?), not the resolution.

replaced by plebiscitary decisions about alternative sets of leaders of administrative personnel. (Habermas, 1971, p. 104).

In this 1971 essay, Habermas apparently viewed this technocratic ideology as highly successful, describing it as "a background ideology that
penetrates into the consciousness of the depoliticized mass ... where it can
take on legitimating power .... Accordingly the culturally defined self-understanding of a social lifeworld is replaced by the self-reification of men
[sic] under categories of purposive–rational action and adaptive behavior."
(Habermas, 1971, pp. 105–106)

However, this relatively strong position does not seem to have lasted
long. In another early essay on the legitimation crisis of late capitalism, the
following statement appeared:

> The political consequences of the authority enjoyed by the scientific system
> are ambivalent. The rise of modern science established a demand for
> discursive justification, and traditionalistic attitudes cannot hold out against
> that demand. On the other hand, short-lived popular syntheses of scientific
> data (which have replaced global interpretations) guarantee the authority of
> science in the abstract. The authority known as "science" can thus cover
> both things: The broadly effective criticism of any prejudice, as well as the
> new esoterics of specialized knowledge and expertise. A self-affirmation of
> the sciences can further a positivistic common sense on the part of the
> depoliticized public. Yet scientism establishes standards by which it can also
> be criticized itself and found guilty of residual dogmatism. (Habermas, 1973,
> pp. 664–665)

I take this to be suggesting that the rise to dominance of science in
modernity may lead to "a positivistic common sense," but it may also give
rise to a more general demand for discursive justification, which can provide
the basis of a critique of the new esoterics. Thus, scientism may lead to
hypnotized ignorance, or to critical enlightenment—or, presumably, to some
combination of both. Nonetheless, it remains an essentially scientific society,
as it is to science that either form of consciousness owes its development.

This is a position, therefore, that is somewhat comparable to Gellner's.
Just as with Gellner, the apparent rejection of science by new social
movements was not really a rejection, because it depended implicity on a
background of acceptance of serious knowledge, so for Habermas, the
critique of science is actually derivative of standards that science itself
sets—and what is really being criticized in any case is not science as such,
but the positivistic and esoteric confinement of science. Thus, science (in
the abstract) is salvaged despite any apparent public rejection.

This basic position is also to be found in *The theory of communicative action*. However, it also has undergone further modification. In particular, it seems that Habermas has tried to reconcile this basic thesis—still pointing to the essential dominance of science in modernity—with the apparent recrudescence of more traditional modes of belief. At least, this is how I interpret his analysis.

The key features here are the discussion of cultural fragmentation and impoverishment (Habermas, 1987b). Fragmentation occurs because of the dissolution of traditional holistic worldviews into their component spheres of the cognitive, the moral–practical, and the aesthetic—this is central to the process of rationalization as Habermas understood it (drawing heavily on Weber; also see Habermas, 1984). However, there is a failure in modernity to reintegrate these spheres at a higher level; this is potentially present, but is frustrated by prevailing powerful interests that benefit from a fragmented social order. Nonetheless, the fragmentation that has occurred is significant. It is a necessary stage in the movement to a higher social order—that is, a fully rationalized (enlightened) one. Further, having occurred, it cannot be reversed, because it implies the structural devaluation of older worldviews:

> With the transition to a new stage the interpretations of the superseded stages are, no matter what their content, *categorially devalued*. It is not this or that reason, but the *kind* of reason, which is no longer convincing. A devaluation of the explanatory and justificatory potentials of entire traditions took place ... in the modern age with the dissolution of religious, cosmological, and metaphysical figures of thought. (Habermas, 1984, p. 68)

However, despite this, because of the inability of modern consciousness to achieve the reintegration that it strives for, it finds itself "thrown back on traditions whose claims to validity have already been suspended; where it does escape the spell of traditionalism, it is hopelessly fragmented" (Habermas, 1987b, p. 355). This is because the newly rationalized spheres of consciousness are not properly available to it: "... the differentiation of science, morality, and art ... results ... in the splitting off of these sectors from a stream of tradition continuing on in everyday practice in a quasi-natural fashion."

In summary, for Habermas, the process of rationalization is one that involves the splitting off of various spheres of a once integrated lifeworld. Science is one of these spheres. In modernity, these spheres have become largely isolated from the rest of the lifeworld—that is, from everyday consciousness—in expert cultures to which most do not have access. In consequence, modern consciousness and culture is fragmented and unable to reintegrate at the higher level of rationality potentially available to it. Instead, it finds itself forced, if at all able, to draw on a stream of quasinatural traditional belief, which provides a pretense of integration, one that is inadequate because the validity of such traditions has been categorially devalued by the shift to a higher sphere. In other words, the only kind of consciousness that can really satisfy the modern is a properly scientific one, but this kind of consciousness is largely unavailable to it. This leads, among other things, to the resort to religious fundamentalism as one form of "the painful manifestations of deprivation in a culturally impoverished and one-sidedly rationalized practice of everyday life" (Habermas, 1987b, p. 395).

However, this position appears to be ambiguous in a way similar to Gellner's. Just as with Gellner, Habermas' analysis seems to demand that modern consciousness as a whole has been affected by the rise of science. For Gellner, this is expressed in the claim that modern consciousness is marked by doubt and skepticism. Habermas' position is very similar: for him, the modern lifeworld is one in which the claims to validity of traditional beliefs—the types of reason given for holding them—are structurally devalued. They cannot therefore sway the modern.

However, just as the question can be asked of Gellner why it should then be that *some* members of modern society appear able and willing to suspend their disbelief and accept outmoded religious fantasies, yet some others see through this, so, too, it can be asked of Habermas why it is that some members of modernity are able and willing to resort to outmoded beliefs, when others (including, presumably, Habermas himself) do not find it necessary to do so.

This problem arises, I think, because of the essential all or nothing quality of the rationalization hypothesis. The claim—of disenchantment, or scientization, or whatever—is made at the highest level of generality; it is supposed to apply to everyone in modernity—and if it does not, it is not clear why it should apply to anyone. Certainly, there are no obvious criteria given for distinguishing those to whom it does apply—those who are, so to speak, really modern—from those to whom it does not and who, therefore, are somehow less than fully modern. This confusion is accentuated by the tendency of these accounts to speak of modern consciousness, or the modern lifeworld in general, as though these entities have an existence of their own, instead of being realized and manifested in particular individuals and social groups. Thus, moving from this level of generality to specific conditions of belief and action becomes fraught with difficulty and ambiguity.

What can be seen in Habermas, then, is an effort to adapt the rationalization hypothesis to a growing sense of cultural complexity, in which the role and position of science has, apparently, been increasingly questioned. To accommodate the presence of such critique seems to require the introduction of divisions into the account, whether involving the nature of science or the relation between science and the wider culture. It is difficult to escape the sense, however, that these divisions are essentially arbitrary, reflecting received academic wisdom and prejudices about the supposed relation between modernity, science, and rationality.

On this basis, it can be suggested that the introduction of such divisions serves rhetorical purposes. They are a further example of ontological gerrymandering (Woolgar & Pawluch, 1985). They provide a means of apparently differentiating out all those things that the theorist wishes to maintain are not really essential to what is being reserved for special treatment. In this case, the special treatment is reserved for science and its supposed impact on modern culture. Science is reserved as a special form of knowledge, by proposing a split between that kind of science that is open to social abuse and that which is not; and yet, the uniformity of its impact on modern culture is maintained by proposing a split between those people who are subject to impoverishment and those who are not.[2]

One of the potential difficulties with this kind of argument, however, is that once you start introducing these kinds of special cases, it can become difficult to know where to stop. Eventually, there are so many exceptions being introduced in order to try to preserve the original rule, that it becomes easier to scrap the lot and start again. This becomes even more tempting if a watershed point in history can be identified, when, suddenly, things have become different. By this means, you can have it both ways: You can keep the original rule and yet scrap it entirely—by suggesting that the original rule applied once, but does no longer. This now common strategy has been given the label postmodernism, but an alternative name for it might be historical gerrymandering.

To illustrate this, we now consider Bell's writings on postindustrialism and cultural contradictions, as this shows a theorist moving from the discourse of rationalization to something rather different. This, in turn,

---

[2]Indeed, Habermas (1987b, p. 383) appeared at one point to suggest that there is within modernity a special group of people, who are more able to recognize the rationality potentials than others—that is, competent members, with an intuitive knowledge of these potentials.

will lead on to consider Lyotard's views of the role of science in the postmodern condition.

Bell

Bell's comments about the social significance of science are to be found in his two major books, *The coming of post-industrial society* (Bell, 1973) and *The cultural contradictions of capitalism* (Bell, 1976) known as CPIS and CCC, respectively. These comments are not always detailed—especially in the second book—but enough can be gleaned from what is said to suggest a clear shift in the perception of the role of science in relation to modern culture between the two books. In CPIS, Bell's position is very much in line with the standard assumptions of the rationalization hypothesis, albeit not without some modification in application. In CCC, however, the impact of science is conceived in a rather different way, with different implications.

In CPIS, Bell also adopted an essentially Mertonian account of science. Science was defined as an ethos of critical skepticism (Bell, 1973). It advances through the communistic sharing of knowledge by a group of equally qualified individuals, each of whom has the right to critically and publicly scrutinize the knowledge-claims of others. Also, again in a similar conception to Merton, he distinguished between the use of science for utilitarian, technical purposes, and its essential nature, which is broadly socialistic, or sociologizing.

Unlike Merton, however, Bell did not think that the ideals of science had been fulfilled within the value system of industrial society. Rather, industrial society is seen as, in effect, a prescientific society, at least in terms of the proper realization of the scientific ethos. Industrial society in defined as dominated by an economizing mode (Bell, 1973), that is, its axial principle is centered on maximizing productivity and is therefore utilitarian, or instrumentally oriented. It is, rather, the postindustrial society that will be the proper scientific society—at least in principle—because this society will be dominated by the sociologizing mode, which characterizes the scientific ethos, especially as realized in the normative expectations of certain key occupational groups, such as professional and other knowledge workers.

Two things can be said, then, about the representation of science in CPIS. First, it is monolithic. Second, its social impact is presented in terms of the dissemination of certain values and normative expectations—the ethos of science—which are realized uniformly in the way in which critical social groups conduct themselves. The values of science are seeded in the conduct of professional workers, who themselves are growing in number and social significance and, as they do so, gradually spreading these values throughout society—again monolithically.

Now, it is true that Bell did not view this as an altogether unopposed process. He identified certain likely points of tension and conflict between the rising scientific professionals and other significant social actors, such as those in the political, industrial, and military sectors who tend to have different value orientations to those of scientists. Also, the ranks of experts are themselves internally divided between the scientists and the artists, the latter adopting a critical, humanitarian stance toward science (cf. Merton, 1968b; Snow, 1964). Even so, science itself remains, in its essential core, untainted—just as it does for Merton and Habermas—and in its core, it is a rationalizing influence.

It is this characteristic of science which seems to evaporate in CCC. In this study, Bell identified what he saw as the contradictory tendencies of modern culture, towards, on the one hand, sober, disciplined rationalism, and, on the other, hedonistic, unconstrained Romanticism—Reason and Will (Bell, 1976). His argument is that we have now reached a difficult period in the development of capitalism, which is increasingly reliant on both tendencies, despite their inherent opposition to each other.

Following Weber, Bell argued that at its outset, capitalism required the ascetic self-sacrifice of the Protestant ethic to provide the basic dynamic of systematic accumulation and reinvestment to get it under way. However, the success of this strategy of production has led, over the course of the 20th century, to the need for increasingly higher levels of consumption. Previously, the urge to consume was largely held in check, confined to expression in artistic movements critical of the sober bourgeois, such as 19th century Romanticism. But over the course of the 20th century, it is to this artistic tradition and the values it embodies that capitalism has increasingly turned to inspire the desire to consume—through, for example, advertising and other aspects of the mass media, which stimulate imaginative fantasy and offer outlets for self-expression. Thus, increasingly, capitalist society has come to demand of its members a split personality: sober by day, swinger by night. Now, in the late 20th century, this has reached a point of exhaustion; hedonism is showing signs of triumphing over rationalism—Will over Reason—but, Bell believed, it is a Will void of content, because it lacks the normative reference needed to give it a stable outlet, that is, it lacks a sense of the sacred.

However, the question of interest here is, where does science fit into all this? It might be considered unfair to ask such a question given that Bell has clearly not set out to discuss the role and position of science. Nonetheless, it can be justified, both in relation to the position outlined in CPIS, and because there is one brief appearance of science in CCC that speaks volumes as far as the current interest is concerned.

Further, the exclusion of science is itself revealing. In CCC, Bell (1976) defined culture as "*a ricorso*," that is, a circle consisting of variations on centrally repeated themes, concerning the existential constants of human life—birth, death, love, tragedy, and so forth. If science were to be included in culture—and Bell (1976, p. 12) did state, in a note, that science, part of the "cognitive mode," is so to be considered—then, presumably, it would also conform to this circle of repetition, in which case, we would lose all semblance of scientific progress. The notion of a gradually unfolding, ever superior understanding of empirical reality would collapse, as would any absolute distinction between science and non-science. Newton would no more replace Aristotle, nor evolution replace creation, than "Boulez ... replace[s] Bach" (Bell, 1976, p. 13). It seems clear, however, that Bell did not want to take such a view of science. Nonetheless, what he had to say about it in CCC reveals further ambiguities.

First, on the basis of the discussion of science in CPIS, we would presumably anticipate it being aligned within the ranks of Reason. Science clearly seems to demand the sober, disciplined self-sacrifice associated with this side of the cultural divide. Bell linked it to the professional ethos of altruistic service, and, within the ethos of the scientific community itself, there is a general valuation of the greater good of the community over the status and ambitions of any one individual. Thus, we would expect the influence of science to be standing against the hedonistic, individualist thrust of consumer culture.

However, in the one passage where science is discussed at any length in CCC, what Bell said suggests otherwise. For it is referred to as among the factors that are responsible for the fragmentation of contemporary culture and the breakdown of a sense of reality and order. This is because science has become increasingly abstract and distanced from the everyday, mundane world, so that it appears more and more to fail in its task of describing this world in concrete terms. Further, when it does talk about this world, what it has to say is that it is not concrete at all! "In a modern cosmology pictures have gone, words have gone ... [leaving] abstract formulas" (Bell, 1976, p. 98). Even these do not refer to hard and immutable laws of nature, but to "uncertainty and the break of temporal and spatial sequence" at the most fundamental, particulate level. Thus, science itself contributes to the "disjunction of cultural discourse" (Bell, 1976: 86) and the incoherence of modern culture—to what many would now call postmodernism (Bell, 1976). A selection of recent texts on this much debated topic includes: Bauman, 1992; Crook, Pakulski, & Waters, 1992; Docherty, 1993; Foster, 1985; Harvey, 1990; Kumar, 1995; Lash, 1990; Lash & Urry, 1987; Lyon, 1994; Lyotard, 1984; Rose, 1991; Smart, 1992; Strinati, 1995).

Thus, whereas in CPIS, science appeared as the source of potential unification of the new social order, in CCC it appeared as one source of disunity. Why?

One possible reason, is that rationalization is a rhetorical construction. It functions as a way of characterizing the modern world that serves particular kinds of discursive purposes. So, for example, if one wants to maintain that modern society is an essentially scientific society, to be demarcated from all other kinds of societies on this basis, then the discourse of rationalization can be drawn on to make the claim. Further, if one wants to maintain this in the face of apparently conflicting characteristics in modernity (or, rather, characteristics that might be said to conflict with it), such as the presence of antiscience movements, then, rationalization provides a set of resources for so doing—such as notions of disenchantment and the reactions it generates (which can be called fragmentation, or impoverishment, for example); or ways of setting up boundaries around what is and is not to be considered a proper social utilization and understanding of science. In this way, the ontology of science and modernity can be gerrymandered to suit the basic argumentative intent, which is to maintain that modern society is a rational and scientific society (even if only in potential).

What Bell's contending views also show, however, is that there are alternative ways of characterizing the social impact of science. What might be suggested about the apparent variation in Bell's position is that he has found himself up against the constraints of the rhetoric of rationalization. Modernity has proved itself too recalcitrant to this analytical solution. Thus, he abandoned this rhetoric and adopted another, utilizing an alternative set of resources, in which the social impact of science is differently characterized, as a culturally destabilizing force, a disruptive influence, rather than a force for stability; and as something which, at least from the point of view of the existing lifeworld (or certain strains within it), may appear irrational, rather than rational.

However, what is not done is to suggest that both of these things may be true; that science may sometimes appear to be rational, sometimes irrational, or indeed both at the same time, depending on who is doing the looking and in what specific circumstances; and that, as members of the modern lifeworld, we might have available to us, both sets of resources for characterizing science, or these and/or a selection of others. Instead, Bell held to the rhetoric that attempts to pin down the nature and impact of science as definitively one thing or another. That this generates enormous confusion can be seen even more clearly in the case of Lyotard, in whose writings the rhetoric of rationalization begins to appear very ragged indeed.

Lyotard

The discussion here will focus on *The condition of postmodernity* (1984). It is useful to consider this text for two reasons. First, although the field of postmodernism almost defies general description, due to the multiplicity of usages of the term, Lyotard is one of the main reference points in these discussions (Rose, 1991). So, his text provides as good a guide as any other—and possibly better than some—to the kinds of thing under discussion and the sorts of assumptions at work.

Second, Lyotard (1984) identified as the defining feature of the postmodern condition, a state of "paralogy," by which he seemed to mean a condition in which there pertains an unlimited agonistics of language games (p. 10). I like this description; it seems to me to fit very well with the rhetorical approach to culture and ideology outlined in chapter 1. Clearly, there is more to be said about this in the final chapter. For now, it is enough to observe the suggestiveness of Lyotard's position and note the possibility of a conceptual meeting. For this to become more than just a suggestion, however, the remaining influence of the rationalization hypothesis in Lyotard needs to be identified and sifted out.

Lyotard's general thesis, somewhat simplified, is that contemporary society is undergoing a major shift due to the collapse of the metanarratives associated with the development of scientific knowledge and understanding. At its core, science is here considered to be a set of procedural principles concerned with the systematic pursuit of validity, that is, the legitimation of knowledge claims. This suspends the traditional mode of cultural reproduction—through orally transmitted narrative—in favor of a process of systematic questioning of assumptions, giving rise to an agonistics of argumentation. The logic of this agonistics, however, has now produced a situation where even the metanarratives themselves have been caught in its questioning light. The reasons for doing science are themselves open to the scientific demand for validation, leading to the complete collapse of their credibility. At the same time, the discoveries of contemporary science have assisted this process at an empirical level, through exposing reality itself as without concrete grounds; wherever we look, whether to the epistemological or to the empirical level, all has now become chaos.

Certain parallels with Habermas and Bell are already apparent. They will become clearer with further elaboration of Lyotard's views of science. Science, claimed Lyotard, has, in the past, been legitimized through two epic stories of its advance: the goal of unified knowledge through a grand synthetic understanding, and the promise of practical empowerment. Both of these metanarratives, however, have now collapsed.

In the first case, the problem is that any proposed synthesis is inevitably open to the same demand of legitimation required of any knowledge claim. However, by virtue of its totalistic character, there are no grounds of validation external to itself to which it can refer. Thus, it inevitably takes the form of a self-legitimating narrative, which, according to Lyotard, is a type of knowledge that science has already rejected in differentiating itself from the earlier cultural form of narrative understanding. Therefore, the hope of grand synthesis is defeated before it begins, as it can never fulfil the demand of legitimation to which it must concede.

In the case of the practical metanarrative, a similar paradoxical difficulty arises. The appeal to practical empowerment engages two kinds of criteria simultaneously: truth and justice, the is and the ought. Practical action needs to be both successful and proper, true and correct. However, the appeal to criteria of propriety reduces science to assessment by standards which are supposedly extrascientific. Once again, therefore, science cannot win: It must satisfy moral grounds of correctness, but if it is open to assessment by such extra-scientific criteria, then it has already lost the claim to be a superior form of knowledge. Thus, this metanarrative also collapses on itself.

These metanarratological developments are paralleled also by developments internal to science, at the level of the formal axioms of the language of science and at the level of empirical proof or observation. Here, one can see a more familiar, Mertonian, distinction emerge, between science as communicative action (or argumentation) and as technical action (Lyotard's performativity). In a similar way to Merton, Habermas, and Bell, Lyotard also saw these two levels of science developing in contrasting directions, albeit for somewhat different reasons.

In the case of argumentation, the "language-game" of science follows a logic of formal deduction, in which propositional statements are defended by reference to first principles and what may or may not be logically deduced from them. However, Lyotard argued that formal logic fails science in this respect, as "all formal systems have internal limitations" (Lyotard, 1984, p. 43). Consequently, science is thrown back on to the resources of natural language. This "allows the formation of paradoxes" (Lyotard, 1984, p. 43), with the upshot that one way in which progress may occur in science is through the invention of a new language-game, giving rise to a heteroglossalia of incommensurable languages. Unlike Habermas, therefore, Lyotard sees the process of argumentation in science as giving birth, not to higher level consensus, but to nonconformity and fragmentation.

This multiplicity and uncertainty, however, contrasts with the other dimension of science, at least in terms of the language-game of performativity. This seems to offer a potential resolution to the crisis of the metanarratives, by legitimizing science through technological development and efficiency maximization. However, this itself is now under threat by the logic of argumentation, which continuously encourages the search for "the unintelligible" (Lyotard, 1984, p. 54), that is, counterexamples that do not fit within the terms of extant language-games. This search has now disclosed a view of reality as chaotic, catastrophic, and probabilistic, which frustrates the deterministic framework of the performative.

It can be seen, then, that Lyotard's analysis bears some similarity to those of Habermas and Bell, albeit his final conclusions, in support of the "paralogy" of the postmodern condition, are distinctive. The similarity resides in his adoption of the language of rationalization, albeit he—fittingly—adopts his own terminology to refer to the processes involved. However, his acceptance of multiplicity brings him into closer accord with the view of science as a cultural resource.

Nonetheless, it is not clear that Lyotard's analysis can be altogether accepted as it stands. Where problems begin to arise, so far as the current perspective is concerned, is in the comparisons he draws between the nature of science and the wider culture, and how he represents the relationship between them. Here, I argue, his analysis is confused and confusing, largely because of his acceptance, albeit in his own uniquely modified form, of the basic thesis of rationalization. There are two key points here: first, Lyotard seemed to characterize science in different ways at different points in his argument, inconsistently—that portrayed a lurking, unresolved dilemma of science; second, these differences articulate with contrary ways of characterizing the public understanding of science, but also help suggest a way in which these problems may be resolved—a way that points toward the view of science as a cultural resource.

Regarding the nature of science, the contrast is clear. As we have seen, Lyotard described science as a type of language-game, which gives rise to a multiplicity of language-games through argumentation; as such, he saw it as offering something of a model or prototype of the postmodern condition, even referring to it as an open system (Lyotard, 1984). However, he also pointed out that the "reality" of science is that it is not always so open: "Countless scientists have seen their "move" ignored or repressed, sometimes for decades, because it too abruptly destabilized the accepted positions .... The stronger the "move," the more likely it is to be denied the minimum consensus, precisely because it changes the rules of the game upon which consensus had been based." (p. 63)

Here we see the same basic contrasting characterization of science as observed in the other discourses and analyses considered in this book. The question of whether science is grounded on open, universal criteria of assessment, or whether it is more local in its determinations, controlled by contingent circumstances of the immediate situation, remains unresolved in Lyotard.

This has implications with respect to the representation of the public understanding of science. Once again, there appears to be something of a contrast here. On the one hand, Lyotard (1984, p. 12) noted a contrast between "the most highly developed industrial societies" and the nature of the general "social bond." This social bond is marked by a general agonistics of language-games, pursued for its own sake. In industrial societies, however, this tendency has been constrained, at least for a period during the 1960s, by the criteria of performativity, which dehumanized (Lyotard, 1984, p. 63) society. The implication of this would seem to be that the wider social and cultural system of modernity is marked by the dominance of this single type of language-game. Similarly, he also referred to the "problems experienced by the social collectivity ... deprived of its narrative culture." (p. 62)

With this characterization of the wider culture, Lyotard can represent science as the van of the new society. This is doubly true, both in that the criteria of performativity are themselves associated with science as one means of legitimation, albeit they have been superseded by developments within science itself; and in that the openness of science to argumentation and new language-games provides the model for the return to the social bond at the new normative capacity of the postmodern condition. Science, therefore, led the way into modern, industrial performativity and it is also leading the way into postmodern, networked paralogy.

However, this relationship between science and the wider culture contradicts a second view found in Lyotard (1984). The wider culture was represented as quite distinct from science, as marked by the presence of narrative knowledge forms, and agonistic in the style of the social bond. Thus, in terms more than a little reminiscent of Habermas, Lyotard (1984) stated:

> Scientific knowledge is ... set apart from the language games that combine to form the social bond. Unlike narrative knowledge, it is no longer a direct and shared component of the bond .... The relation between knowledge and society (that is, the sum total of partners in the general agonistics, excluding scientists in their professional capacity) becomes one of mutual exteriority. (p. 25)

Putting aside the rather schizoid image of scientists this seems to convey between their everyday and professional personae, this second view seems rather at odds with the first representation of science in the van of modernity. Similarly, Lyotard (1984) stated:

> Th[e] return of the narrative in the non-narrative ... should not be thought of as having been superseded once and for all. A crude proof of this: what do scientists do when they appear on television or are interviewed in the newspapers after making a "discovery"? They recount an epic of knowledge that is in fact wholly unepic. They play by the rules of the narrative game; its influence remains considerable not only on the users of the media, but also on the scientist's sentiments. This fact is neither trivial nor accessory: it concerns the relationship of scientific knowledge to "popular" knowledge, or what is left of it. ... [T]he State's own credibility is based on that epic, which it uses to obtain the public consent its decision makers need. (pp. 27–28)

Despite the qualifications here—and leaving aside the extent to which this notion of the epic storying of science is actually true or not[3]—it is difficult to reconcile this view with the first view of industrial life as dominated by performativity. Why would those constrained to the language of performativity afford any credibility to the narrative form of the epic? Again, then, there seem to be contrasting representations advanced at different points in the analysis, contrasts which expose the unresolved

---

[3]The analysis of media representations of science is an emerging field of research in its own right. For an excellent summary, see Lewenstein (1995).

difficulties over the nature of the public understanding of science within the rationalization thesis.

To conclude this, Lyotard (1984) sometimes appeared to continue to describe modern society in terms of the general social bond. Thus: "Social pragmatics does not have the 'simplicity' of scientific pragmatics. It is a monster formed by the interweaving of various networks of heteromorphous classes of utterances (denotative, prescriptive, performative, technical, evaluative, etc.)." (p. 65)

In similar vein, Lyotard said that "popular narrative pragmatics" and "the language game known to the West as the question of legitimacy" are incommensurable. Consequently, narrative knowledge's "incomprehension of the problems of scientific discourse is accompanied by a certain tolerance: it approaches such discourse primarily as a variant in the family of narrative cultures." (Lyotard, 1984, p. 27)

This seems to add up to a view of the relationship between science and the wider social and cultural order as one of complete separation, rather than, as the earlier view seemed to suggest, one in which science is guiding the whole system towards the postmodern. Indeed, from this latter quotation, one might even take the view that it is narrative that presages the postmodern condition rather more than science. Add in Lyotard's own observation that science is not as open as it is often claimed to be and this inference becomes more persuasive still. However, to pursue this further, it is first necessary to make some adjustments to Lyotard's position, to straighten out the confusion. This essentially means abandoning the remaining elements of the rhetoric of rationalization that bedevil his account.

From Rationalization to Resource

The discussion of these three theorists demonstrates the presence of a shared discourse of science within their accounts of modern society. This discourse articulates around two apparently conflicting versions of science, which give rise to systematic variations and symptomatic ambiguities in the representations of its role and position in modernity. These variations and ambiguities are characteristic of the rationalization hypothesis.

This has been seen in each theorist in turn. In Habermas, there is an unresolved difficulty over the status of science (especially natural science) as to whether it is a technical or a communicative process. From an earlier position that appeared to exclude science from the realm of the communicative, he has moved to one in which he seemingly ascribes a dual status to science as both technical and communicative. The viability of this position is doubtful, however, as it seems inevitable that, once matters of empirical truth are included within the arena of the argumentative pursuit of validity, it is then unclear how they can ever be said to be properly uncoupled. Once the argument has started, why should it ever stop?

In Bell, a similar difficulty arises, although it is modified by his adoption of a different strategy to resolve the dilemma that arises with rationalization. In his case, rather than attempt to split science into two, he split society itself, equating the sociologizing mode of science with the coming postindustrial order and treating the economizing mode as essentially prescientific. However, this position, too, becomes problematic, as is shown by his exclusion of science from the analysis of cultural contradictions. To include science would dissolve the boundary erected by rationalization between science and nonscience, between the argument that progressively advances toward consensus (because more true), and the argument that is destined endlessly to recycle itself with neither advance nor truth in sight.

Even Lyotard, who seems to offer a view of science as communicative multiplicity, nonetheless still seems to insist also on trying to construct a fence around it and reserve for it a special status. This merely accentuates the ambiguity. Is science the model of the postmodern condition, the advanced prototype of the emerging state of paralogy, in which multiple language-games contend in an endless play of invention? Or is it really quite open enough to be up to the job? And if not, why choose it as the model? Again, is it the wider culture of modernity that is dominated by the discourse, of performativity? Or is this discourse itself merely one of the many language-games that constitute the heteromorphous monster of the social bond? In Lyotard, these questions come to a sharp head, precisely because the shifts and movements in his argument, and the range of versions of science within his writings are so manifest. What this points to is a need for radical overhaul, specifically to wipe away the rhetoric of rationalization with its representation of science as an asocial monolith, dominating the social and cultural order of modernity. The monolith is unstable and its rhetoric no longer convincing.

What, then, is to replace it? Rationalization seeks to resolve the central dilemma of science by postulating modernity to be a society in which social action itself is becoming more purposive (Weber), technical (Habermas), economizing (Bell), or performative (Lyotard). Within this framework, science is represented as one embodiment of this kind of action. However, to accept this position as it stands generates difficulties in making sense of the public understanding of science, which is marked by the continuing presence of various antiscientific anomalies. Rationalization has to see these movements as anomalous, because science is taken to be essentially monolithic. There is only one science; there can be only one, because science is simply a rational, technical reflection or embodiment of a single empirical reality. How, then, are antiscientific movements in an essentially scientific social context to be understood?

The various strategies adopted by the theorists reviewed at various points in this book are designed to solve this problem. One strategy is to split culture, but this merely defers the problem, as it does not explain how it is that the antiscientific and irrational elements of culture can persist at all. A second strategy is to split science itself into technical and communicative aspects. Yet, this exposes to question the basic assumption of the essential unity of science within rationalization, with two implications. First, it generates a boundary problem between the technical and communicative aspects, specifically between the empirical determination of knowledge and its determination by consensual agreement arrived at through argumentation. Second, this generates further ambiguities with respect to the public understanding of science, as the designation of this understanding in particular cases as either acceptable (because in accord with the communicative logic of science) or unacceptable (because impoverished, or misunderstood, or merely playful, or whatever) is essentially arbitrary.

A third strategy is to split society itself into technical (i.e., prescientific) and communicative (i.e., scientific) forms, such as the industrial–modern and postindustrial–postmodern. However, this strategy merely shifts the same boundary problems to a more generalized and abstracted level, contributing further to the sense of arbitrariness and ambiguity. It substitutes for detailed analytical understanding a game of categorization in the designation of things as modern or postmodern, arguably contributing little more than the multiplication of theoretical versions (see Rose, 1991). Additionally, in the case of both science and the public understanding of science, it is difficult to see how this move can resolve the boundary problems satisfactorily, as it merely seems to restate them in a sharper form.

At this point, one might be inclined to make an appeal to still one more model of science, that of Ockham. The complexities of theorization that the attempt to operate within the terms of rationalization discourse generates are just too great. Something simpler is called for. This can come from abandoning rationalization and, in particular, its assumptions about the unity of science and the public understanding of science. Further, there is no difficulty with doing this. As discussed in chapter 1, there is an alternative view of science readily available from SSK and DA in particular—a view that does not postulate the analytical abstraction of a unified science, but instead views science as a form of social action akin to any other. Unlike rationalization, it does not try to reserve for science a special status as a pure kind of technical action. Rather, it sees it as simply another form of social, or communicative action. By the same token, SSK should view the public understanding of science in the same way: as simply another form of social action, marked by the same basic features. It is this view that is advanced in Barnes' notion of science as a cultural resource, to be further elaborated with reference to creationism in the final chapter.

7

Rhetoric as Usual

I have tried to argue a number of things in this book, and a brief review would seem to be in order before any further attempt to address their implications for understanding the role and position of science in modernity. Thus, before moving on to discuss further the view of science as a cultural resource and its agonistic relations with alternative fields of discourse (such as Christianity), I now summarize the main lines of argument of the previous chapters.

1. There is a shared discourse of science in modernity, characterized by contrasting representations of it as, on the one hand, a monolithic, asocial, universal category, and on the other, as a local, human, and socially grounded, form of knowledge or belief. This discourse is worked out in the form of a repertoire of models of science, ranging across these contrasting representations, and in modes of argumentation designed to support or undermine particular views of the nature of reality. Two main modes of argumentation stand out: one, the empiricist repertoire, draws on an impersonal, asocial and universalistic rhetoric to claim that it speaks the way the world is in its unmediated self; the other, the contingent repertoire, draws on a personalized, human and socially grounded rhetoric to undermine such universal claims, by referring them to the particularities of the context of their production. A number of studies now have shown that these rhetorics and repertoires are used by both scientists and nonscientists alike and the claim here is that they provide a central resource for contemporary so-called antiscience movements in mounting their critical attacks on the orthodox scientific community (Locke, 1997).

2. Support for this comes from the case of creationism—and from its sociological critics. A further argument of this book is that both of these groups draw on similar representations and rhetorics of science. This is seen especially in their use of different models of science to construct their accounts. Creationists, concerned to account for evolution, draw on a range of models of science in order to undermine its legitimacy, credibility, and validity, and to advance the case for creation; sociologists, concerned to account for

197

creationism, also draw on a range of models of science in order to undermine its (creationism's) validity and thereby justify the necessity and validity of their sociological views. Thus, the two groups share a pattern of argumentation, one that draws on the repertoire of models of science as means of legitimation and instrument of persuasion. In so doing, they also construct their opponent, and/or the phenomenon to be explained, in a manner that suits the account being advanced. For example, a creationist may draw from the repertoire of models of science a version of falsificationism, then construct a view of evolution as nonfalsifiable, and so undermine its claim to scientificity; or the same version of science may be used, with a construction of evolution as falsifiable, then to claim that it has been falsified. Similarly, a sociologist may claim that creationists view science in inductionist terms, then advance an alternative version of science—as, say, conventionalist—so undermining the credibility of creationism; or creationism may be presented as essentially moral, with science presented as essentially empirical, again undermining its validity. And so on.

3. A particularly common construction of creationism in sociological texts is as a new phenomenon. A third argument in this book is that this is an outcome of the discourse of rationalization, characteristic of sociological understandings of the role and position of science in modernity. The rationalization hypothesis represents science as a monolith, marked especially by the dominance of a technical mode of action, which spreads throughout the wider society, displacing communicative forms of action, through a logic of disenchantment. Although details are a little unclear, this process seems to have reached some kind of significant watershed in the aftermath of the Second World War, with the establishment of science as pivotal in an increasingly centralized and bureaucratic state and industrial order. However, it has also left a prevailing sense of meaninglessness, a moral vacuum further exacerbated by alienation from the center. This has provoked a search for alternatives to science, drawing on localized, traditional modes of belief, including such things as Christianity. However, although widespread in sociology, the discourse of rationalization is unable adequately to account for the puzzle of creation science. Rationalization represents modernity as a scientized society; public criticism of science, therefore, generates confusion in the account. It has to be understood in some way as not real criticism, either because it is really only superficial playing and not serious, or because it misunderstands science and so is directed at the wrong target.

Such accounts, however, only defer the problem; the source of the playing or misunderstanding still needs to be accounted for. Moreover, they do not allow for the possibility of directed critique and selective rejection of science, displaying varying forms of understanding and degrees of argumentative sophistication. The crisis of the discourse of rationalization prompted, in part, by the effort to address contemporary phenomena of this nature becomes more apparent from examination of the representations of science and the public understanding of science in the writings of some recent social theorists, notably Habermas, Bell, and Lyotard. What this shows particularly is that an alternative representation of science—as communicative rather than

technical—is drawn on, but not reconciled with the monolithic representations of rationalization. The struggle to resolve these problems seems
to produce increasingly stark ambiguities and increasingly extreme divisions—within the conceptions of science, culture, and society—which seem
to betray a sense of theoretical frustration and failure.

4. As an alternative to this, a further argument in this book is in defense of the
idea of science as a cultural resource. Originally advanced by Barnes as a
corollary of his early statement of the need for and possibilities of a sociology
of scientific knowledge, this is a view of science as a members' category, with
its meaning to be discovered from its utilization within specific social contexts, rather than imposed from outside by analytical fiat. Further, whereas
rationalization assumes a monolithic science with a singular, developmental
logic spreading throughout society, provoking disenchanted responses,
science as a cultural resource assumes multiplicity and variation. It anticipates versions of science; thus, it predicts that the boundary between
science and nonscience will be moveable and more specifically, a construction produced as an outcome of processes of interpretation, explanation and
argumentation within specific contexts.

This provides a means of resolving the confusions of rationalization. These
confusions arise because of the assumption of a monolithic science and the
attempt by analysts to impose this on a diverse and fragmented social order.
Further, there is no need to assume that this is a novel state of affairs—a
postmodern or any other such condition; rather, multiplicity is the normal
state of modernity. As Gilkey suggested in his account of creationism (once
cleansed of the infecting rhetoric of rationalization), science is something that
is shaped at every level by the diverse social groupings of modernity, and it
is this that makes science what it is, at least as far as the public understanding
of science is concerned. Similarly, as Wynne said, public understandings of
science can be expected to be multiplex, often perhaps seeming contradictory, because different versions of science serve different purposes for
different individuals and groups in different social contexts.

This position encourages a focus on the processes of construction of
science in public contexts—that is, on the rhetorics of its representation and
the forms and modes of argumentation used in defense and attack. Added
to this, it is no longer necessary to assume, as is the case with rationalization,
that alternative discourses to science are necessarily outmoded or displaced
by its advance. This only follows from the assumption that science portrays
the truth of reality in a direct and singular fashion, such that it can neither be
doubted nor questioned, nor accounted for in any other way (unless irration-
ally). From the point of view of science as a cultural resource, however, this
representation of science must be seen as only one possible representation
and therefore, among wider publics, there may well be other views taken of
science and the knowledge claims made by scientists, views that might also
draw upon other kinds of cultural resource. These cultural resources might
be seen by at least some as alternative sources of authority to science,
providing means of resistance and opposition, in the form of alternative
knowledge claims and constructions of reality, and the rhetorical instruments

with which to construct a critical response. Or, they may contribute to a growing pool of cultural resources, increasingly widely available, perhaps, from which larger numbers of individuals and groups are able to draw to construct versions of science and versions of other things, in a growing diversity of exotic syncretisms.

5. This is what has been shown here in the case of creationism. Thus, a further argument of this book is that creationism—and, I would suggest, other forms of so-called antiscience—is better understood, not so much as a response to a monolithic science imposed from above, but as an active constructor of the meaning of science, at least in certain social contexts and for certain social groups. And, of course, there is not just one meaning of science within creationism itself; rather, it is part and parcel of the rhetorical processes in which creationists engage to construct multiple versions of science, representing it in different ways at different times in order to achieve different argumentative outcomes. Their rhetoric in this respect is designed to serve one particular purpose: to undermine evolution and bolster creation in its stead. To do this, they draw on different models of science and on the broader intepretive repertoires of empiricism and contingency. In this way, their procedures of argumentation are designed to manage the existence of evolution, to delegitimize it and account for the errors of its way. And, as said previously, in this respect, creationists' discourse parallels the discourse of scientists and of sociologists attempting to account for creationism. This points to the presence of a shared discourse of science in modernity, a discourse that is itself marked by internal tension, or a dilemma.

6. Thus, the final argument of this book is that science is dilemmatical. The dilemma emerges because science makes claim to universal validity, but it is a local activity, grounded in the particular actions of particular people in particular contexts. It is this contrast that gives rise to the shared discourse of science in modernity, defined by the repertoire of models of science and the complementary argumentative voices of empiricism and contingency. As a consequence, it is this internal dilemma that itself provides the resources for the critical deconstruction of science, such as that undertaken by creationists. It should come as no surprise, then, that there are many other such deconstructions to be found within modernity. To consider this a little further, I conclude, with a brief discussion of the notion of ideological dilemmas.

IDEOLOGICAL DILEMMAS OR RHETORIC AS USUAL

Central to the view of rhetoric taken in this book is the idea of competing *logoi*. I take my understanding of this from Billig (1987, 1991), who drew his inspiration from Protagoras. Billig highlighted a contrast between Protagoras and Aristotle, with regard to the nature of things. For Aristotle, things cannot be both what they are and what they are not; for Protagoras, however, "on every issue it is possible for it to be argued with equal force on both sides" (Billig, 1991, p. 24). So, things can be both what they are and what they are not; they can be made to be more than thing, to be many things, to be contrasting things, all by the artful use of argument—what Billig (1987) called *witcraft*. With witcraft, competing views of the world—different *logoi*—may be defended equally well. Moreover, they may be defended using similar types of argument—for and against. Traditionally, the study of rhetoric was, in part, the study of the different kinds of argument and the general characteristics of their working. To become skilled in rhetoric, then, was to be able to mobilize the same kinds of argument to support different, indeed, opposing views. This is at least part of the reason why rhetoric came to be viewed pejoratively.

In these terms, evolution and creationism—or, science and Christianity—can be seen as competing *logoi*, presenting different versions of the world, but using much the same argumentative modes and techniques. They may also carry equal force as well, at least for some. Following Barnes' (1974) view of the contextual nature of explanation, the force, or value of an argument—and, therefore, its persuasiveness—is similarly determined contextually (cf. Antaki, 1994). To argue, as Habermas (1984) did, that certain kinds of beliefs have been "structurally devalued," imposes an artificial uniformity on society—an acontextual consensus that actually tries to dodge the argument by default (some things are not worth arguing with) and fails to recognize that the need to persuade never ends. Moreover, it is not clear that, assuming it were achievable, such a structurally ingrained consensus is compatible with a truly emancipated society; as Nelkin (1982) argued, evolutionists often resort to their own form of fundamentalism in their hostility to creationism. Dismissing opposing positions as irrational or as misunderstanding can appear to be making excuses for not arguing and, perversely, used to bolster the opponent's position. Moreover, it is likely to prompt similar responses in return. Worse still, it is poor rhetoric.[1]

As already suggested, Billig's Protagorean viewpoint also provides grounds for proposing that the case of creationism may be only one instance

---

[1] As McPhail (1996) said in discussing two pieces of writing with which he found himself in strong disagreement:

> both challenged me to reflect *critically* on my own beliefs concerning what it meant to be "open minded". [....] I could easily judge either of these two individuals in terms of *their* ignorance, of *their* not knowing what *they* were talking about.... [But] I knew that to do so would be to knowingly and willingly violate the rule of justice, to reaffirm my own smug sense of superiority; I knew that to do so would make me, *essentially*, no different from those I would condemn. (p. 35)

of a more general phenomenon. To generalize too readily, however, would not be advisable, as there are also good reasons for suggesting that creationism may be somewhat distinctive. Unlike, perhaps much of the stuff of everyday commonsense, creationists are relatively well-organized and systematic. It is likely, therefore, that their arguments will be that much more fully developed and established. They also have a strong tradition behind them, with significant grassroots support. Also, their alternative discursive resource—the Bible—is widely known and readily available. By the same token, however, they also have to contend with equally well organized and established opposition from within the Christian tradition, as well as from orthodox science—although, from the rhetorical point of view, a stance may well become more effectively developed when it has a sturdy, well-trained opponent, pushing it to produce an even sturdier defense.

Acknowledging these qualifications, I nonetheless want to suggest, in the spirit of scientific (!) conjecture, that there is a more general phenomenon at work, concerning the relation between science and other possible *logoi* that might be available within modern culture. Here, again, what Billig (1987) said about commonsense is suggestive. Commonsense can be thought of as a repository of arguments, often in the form of saws and sayings, clichés, maxims, and so on. Often, these can seem to offer conflicting advice, as in the (not altogether inappropriate) contrast: He who hesitates is lost, but look before you leap. Such conflicts, however, provide food for thought and, more importantly, both the grounds and the means of argumentation. In a word, they pose dilemmas. Now, the relationship between such everyday saws and sayings and grand ideological constructions such as science and Christianity is by no means direct or obvious. Nonetheless, the view has been advanced that, as much as the folk-wisdom contained in saws may be contrary, so, too, may ideologies as a whole.

This was suggested by Billig et al. (1988). They proposed that it is the internal dilemmas of ideologies that are at the center of their everyday practical implications and usage. They used the term *ideology* in its broad structuralist sense to refer to the prevailing general outlook, mode(s) of belief and means of identity formation within a society or culture. For my purposes, however, I refer to this level as the commonsense lifeworld and reserve the term ideology for specific discursive fields, such as science and Christianity. What I want to take from Billig et al., nonetheless, is the claim that ideologies (at whichever level) are not internally coherent (cf. Gilbert & Mulkay, 1983); rather they are dilemmatical, that is, offering contrasting judgments and modes of representation, with associated differences in what they imply for making sense of situations and dealing with them through specific courses of action. Such dilemmas provide actors with their food for thought, as in all situations they are or may be confronted by choices over how to act and what to think, choices that demand that they make up their own minds for themselves.

Of course, they (we) do not do so in complete isolation. We decide what to think, in part, through discussion with others—that is, through argument. Herein resides the connection between the notion of ideological dilemmas and Billig's rhetorical view of thinking. What seems to be suggested is that dilemmas have a twofold function: They provide the background context against or within which thought and action takes place and, as such, they provide a resource for constructing understandings of actions and situations; these, in turn, lead to further thinking about the next series of actions to follow in the new situation that will have developed, which will involve renewed application of the same dilemmatical resources; and so on. Thus, because these resources are dilemmatical, alternative interpretations of situations and accounts of past actions are always possible; in consequence, actors will find themselves able to present, and/or being presented with, alternative case scenarios. This may well give rise to argument, or agonistics. Argumentation, therefore, is normal and it is no surprise then to find normative ritualization and standardization of the forms and modes of argumentation, such as in the saws and sayings central to the commonsense lifeworld.

In this, it is possible to begin to see some connection between the dilemmatics of ideologies, as advanced by Billig et al., and a Habermasian view of the lifeworld as marked by communicative rationality, itself understood as the raising and testing of validity claims through procedures of argumentation. Of special note is the insistence of Billig et al. that ideologies, dilemmas and arguments are not merely the province of specialized experts. Rather, the stuff of everyday thought and action consists of the same basic dilemmas and arguments that intellectuals trouble themselves over. Such dilemmas may be readily identified in the lifeworld of modernity, for example, in the dilemma over science and religion.

The reason I wish to distinguish between ideologies, or discursive fields, and the lifeworld, is because I find Billig et al.'s position in this respect too constraining. The problem here is that their focus is solely on the ideology of the Enlightenment and the dilemmas this gives rise to in everyday contexts—dilemmas, for example, between freedom and determinism, and equality and authority. The suggestion seems to be that they see the ideology of the Enlightenment as coterminous with the commonsense of modernity.

If this is the case, I think it is mistaken; certainly, to take such a view would make the interpretation of creation science, in its full syncretic characteristics, difficult.

My proposal is that Enlightenment thought constitutes only one of the ideologies—discursive fields—to be found within modernity. There are others also, such as Christianity. This is not to suggest that the discourse(s) of the Enlightenment has not been extremely successful—that is, persuasive to the modern mind. It is, however, to suggest that the degree of this persuasiveness must itself always be seen as contextual, and it is to suggest also that an important factor in limiting the success of Enlightenment ideology is likely to be its own internal dilemmatics.

This is what I have been arguing throughout is the case with science and its relation to religion. I have argued that Enlightenment discourse of science is itself marked by an internal dilemma, between the presentation of science as a universal form of knowledge, and its presentation as a product of local circumstances. This not only stimulates argumentation within the discursive field of science as to its nature and implications, but also between this field and those of other ideologies. This can be seen in the case of Christianity, with which science has been involved in a continuous, if not unchanging, argument since at least the early modern period (as it has also with occultism—see Yates, 1972). However, Christianity is itself also internally fragmented and dilemmatical. Consequently, a range of positions has been established within this discourse, just as is the case within science. Equally, a variety of possible relations between these two fields of discourse has emerged, ranging from versions of compatibility and incompatibility, mutuality and exteriority, and all points in between (cf. Barker, 1980). These, in turn, have stimulated further levels and layers of argument.

Critically important to recognize, however, is that these positions and the dilemmas from which they arise, are not necessarily confined to academics, intellectuals, and professional scientists and theologians. Rather, they are the very stuff of the lifeworld and its agonistics. Lyotard borrowed from Wittgenstein a delightful image to describe this agonistics, which although applied to language, works equally well as a description of the modern lifeworld: "Our language can be seen as an ancient city: a maze of little streets and squares, of old and new houses with additions from various periods; and this surrounded by a multitude of new boroughs with straight regular streets and uniform houses." (Wittgenstein, cited in Lyotard, 1984, p. 40) As Lyotard (1984) added: "New languages are added to the old ones, forming suburbs of the old town .... " Many such suburbs have been added over the past few hundred years by the multiplicity of languages wrought by the sciences; some might even want to claim that whole new gleaming cities have been built. This does not mean, though, that the older parts of town have been completely abandoned or forgotten, nor that they might not become swiftly repopulated, perhaps with a new coat of paint, or an extension added with materials brought from the new suburbs.

Such is creation science. Creation science is a construction built from the discursive resources of both old and new languages and, as such, shows the workings of dilemmas between ideologies. On one level, creationists can be understood as trying to resolve the dilemma posed by not wanting to reject belief in the Bible as the true Word of God, nor wanting to reject outright the evidence from the world that is taken by the scientific orthodoxy as disproving Genesis. They want to satisfy both the Word of God and the Word of the World. To achieve a resolution of this dilemma between ideologies, they exploit the internal dilemmas within them, constructing a position which, they claim, satisfies both sides. Thus, they exploit the dilemma of science between the universal and the local in both their use of the repertoire of models of science and the conjunction of empiricist and contingent rhetorics. Similarly, they exploit the dilemma of Christianity over the nature of a literal reading of the Bible. They use the rhetoric of fundamentalism to present an interpretation of the biblical text that denies its own interpretive status, defending their position with the conjunction of the rhetorics of fundamentalism and contingency. Thus, they are able to construct compatibility between their version of science and their version of the Bible.

It is, then, the existence of such internal dilemmas within the ideologies of science and Christianity that makes creation science possible. Creationists use the internal dilemmas to provide them with arguments in support of their central concern, to resolve the external dilemma between the ideologies. Their position, therefore, should be understood as a rhetorically achieved dilemmatical resolution, or, as I have called it, a discursive syncretism.

There is no particular reason to assume that creationism is unique in this respect. Modern culture is not marked by just one ideology, or discursive field, but by many. This is apparent from the accounts of both Habermas and Lyotard. Our town is old; it has many mansions and many alternative building blocks, in the form of the dilemmas and arguments provided by a range of different ideologies.

Indeed, it is one of the strengths of the postmodern thesis that it acknowledges this characteristic. Where postmodernism is surely mistaken, however, is in the assumption that modern culture has ever been any different than this. Surely it has always been a multiplicity. Here Billig et al.'s thesis is persuasive. In seeing ideologies as always marked by dilemmas, and thought as marked by arguments, they are recognizing that coherence and/or consensus is, as Lyotard also observes, merely a momentary stopping place before the argument begins again. But this is not a postmodern condition; it is simply a case of rhetoric as usual.

And, as I have suggested, it is only a short step from here to propose that the lifeworld of modernity is argumentative, not just because of the dilemmas contained in one ideology, but also because of those between ideologies, such as that between science and religion, which might, in one form, also be expressed as between the material (or "this-worldly") and the spiritual (or "other-worldly"). It is not too difficult to think of examples of this general kind of dilemma informing the commonsense lifeworld of contemporary industrial societies, in debates over abortion, brain-death, euthanasia, genetic engineering, nuclear power, and so on. New Age beliefs, much as they may be derided in some quarters, can also be seen as in part addressing similar dilemmas,[2] as indeed, arguably can many of the movements and pop philosophies that comprise the "cultic milieu" (Campbell, 1972) in general.

A further valuable study that lends some support to the present thesis is Hess' (1993) examination of the shared "paraculture" of science amongst "New Agers," researchers of the paranormal, and the self-proclaimed skeptics of the Committee for Scientific Investigation of Claims of the Paranormal (CSICOP).

More research is needed to explore these applications in detail, but enough has been said already, I hope, to give at least some credence to the suggestion that these movements may be understood as phenomena of the general argumentative agonistics of modernity. Within this agonistics, religion is far from being merely a dead tradition, however much its institutional position may have declined (Wilson, 1966). Indeed, the decline of this position is likely to have contributed as much to the general capacity to debate religious and moral concerns from a multiplicity of viewpoints as it has to anything that might be called secularization. Equally, the development of materialistic science should be seen as merely adding further dimensions to this process of collective argumentation. It is not a matter of

---

[2]Restivo's (1978, 1982) excellent analysis of Capra's *Tao of Physics* (1975) deserves mention here. Although he offers a rather different account of the significance of Capra, nonetheless, he showed clearly the manner in which Capra's synthesis of modern physics and Taoism rests on a highly selective understanding of both types of knowledge. In my terms, based on Restivo's account, Capra constructed a compatibility between them, providing a new discursive syncretism.

a simple one-way dominance in which science, whether understood as critical or as positivistically impoverished, dominates all else. Rather, science is taken up in the spirit of argument. This spirit already exists in the wider culture, because it is inherent in the dilemmatics of ideologies and the rhetorical agonistics that these engender. Within this, what is and is not to be considered science, and how science is to be understood, itself becomes part of the debate. There is no simple contrast between the scientific and the unscientific, but a continuum of rhetorics providing a range of discursive resources, each of which may be drawn on to varying degrees and in varying contexts, producing a kaleidoscope of models, metaphors, language-games, and validity claims continuously aligned and realigned by the agonistics of our dilemmatic communicative rationality.

That, at any rate, is *my* argument, my *bricolage*.

REFERENCES

Ang, I. (1996). *Living room wars: Rethinking media audiences for a postmodern world*. London: Routledge.

Antaki, C. (1994). *Explaining and arguing: The social organization of accounts*. London: Sage.

Ash, B. (Ed.). (1977). *The visual encyclopedia of science fiction*. London: Pan.

Ashmore, M. (1989). *The reflexive thesis: Wrighting sociology of scientific knowledge*. Chicago: University of Chicago Press.

Ashmore, M., Mulkay, M., & Pinch, T. (1989). *Health and efficiency: A sociology of health economics*. Milton Keynes: Open University Press.

Ashworth, C. E. (1980). Flying saucers, spoon-bending and Atlantis: A structural analysis of new mythologies. *Sociological Review, 28,* 353–376.

Atkinson, J. M., & Heritage, J. (1984). *Structures of social action: Studies in conversation analysis*. Cambridge, England: Cambridge University Press.

Barker, E. (1979). In the beginning: The battle of creationist science against evolutionism. In R. Wallis (Ed.), *On the margins of science: The social construction of rejected knowledge* (pp. 179–200). Keele: University of Keele Press.

Barker, E. (1980). Science and theology: Diverse resolutions of an interdisciplinary gap by the new priesthood of science. *Interdisciplinary Science Reviews, 5,* 281–291.

Barker, E. (1985). Let there be light: Scientific creationism in the 20th century. In J. Durant (Ed.), *Darwinism & divinity: Essays in evolution and religious belief* (pp. 181–204). Oxford, England: Basil Blackwell.

Barker, J. (1996, November). *In the beginning: Christian and Aboriginal creationisms in Canada*. Paper presented at the American Anthropological Association, 95th Annual Meeting, San Francisco, CA.

Barker, M. (1989). *Comics: Ideology, power and the critics*. Manchester, England: Manchester University Press.

Barnes, B. (1974). *Scientific knowledge and sociological theory*. London: Routledge & Kegan Paul.

Barnes, B., Bloor, D., & Henry, J. (1996). *Scientific knowledge: A sociological analysis*. London: Athlone.

Barnes, B., & Edge, D. (Eds.). (1982). *Science in context: Readings in the sociology of science*. Milton Keynes, England: Open University Press.

Bauer, M., & Schoon, I. (1993). Mapping variety in public understanding of science. *Public Understanding of Science, 2,* 141–155.

Bauman, Z. (1992). *Intimations of postmodernity*. London: Routledge.

Beck, U. (1992). *Risk society: Towards a new modernity*. (M. Ritter, Trans.). London: Sage.

Bell, D. (1973). *The coming of post-industrial society: A venture in social forecasting*. London: Heinemann.

Bell, D. (1976). *The cultural contradictions of capitalism*. London: Heinemann.

Berger, P., & Luckmann, T. (1971). *The social construction of reality: A treatise in the sociology of knowledge*. Harmondsworth, England: Penguin.

Bernstein, R. J. (1976). *The restructuring of social and political theory*. Oxford, England: Basil Blackwell.

209

Billig, M. (1985). Prejudice, categorization and particularization: From a perceptual to a rhetorical approach. *European Journal of Social Psychology, 15*, 79–103.

Billig, M. (1987). *Arguing and thinking: A rhetorical approach to social psychology.* Cambridge, England: Cambridge University Press.

Billig, M. (1991). *Ideology and opinions: Studies in rhetorical psychology.* London: Sage.

Billig, M., Condor, S., Edwards, D., Gane, M., Middleton, D., & Radley, A. (1988). *Ideological dilemmas: A social psychology of everyday thinking.* London: Sage.

Bittner, E. (1974). The concept of organization. In R. Turner (Ed.), *Ethnomethodology: Selected readings* (pp. 69–81). Harmondsworth, England: Penguin.

Bloom, W. (Ed.). (1991). *The New Age: An anthology of essential writings.* London: Rider.

Bloor, D. (1976). *Knowledge and social imagery.* London: Routledge & Kegan Paul.

Boden, D. (1994). *The business of talk: Organizations in action.* Cambridge, England: Polity; Oxford, England: Blackwell.

Boden, D., & Zimmerman, D. H. (Eds.). (1991). *Talk and social structure: Studies in ethnomethodology and conversation analysis.* Cambridge, England: Polity; Oxford, England: Blackwell.

Brubaker, R. (1984). *The limits of rationality: An essay on the social and moral thought of Max Weber.* London: Allen & Unwin.

Bruce, S. (1984). *Firm in the faith.* Aldershot, England: Gower.

Bruce, S. (1988). *The rise and fall of the New Christian Right: Conservative Protestant politics in America 1978–1988.* Oxford, England: Clarendon.

Burman, E., & Parker, I. (Eds.). (1993). *Discourse analytic research: Repertoires and readings of texts in action.* London: Routledge.

Cameron, I., & Edge, D. (1979). *Scientific images and their social uses: An introduction to the concept of scientism.* London: Butterworths.

Campbell, C. (1972). The cult, the cultic milieu and secularisation. In M. Hill (Ed.), *A sociological yearbook of religion in Britain 5* (pp. 119–136). London: SCM Press.

Campbell, J. A., & Benson, K. R. (1996). The rhetorical turn in science studies. *Quarterly Journal of Speech, 82*, 74–109.

Capra, F. (1975). *The Tao of Physics.* London: Wildwood House.

Cavanaugh, M. (1985). Scientific creationism and rationality. *Nature, 315*, 185–189.

Chalmers, A. F. (1982). *What is this thing called science? An assessment of the nature and status of science and its methods* (2nd ed.). Milton Keynes, England: Open University Press.

Clute, J., & Nicholls, P. (1995). (Eds.). *The encyclopedia of science fiction.* London: Orbit.

Coleman, S. M., & Carlin, L. E. (1996, November). *Opposition or accommodation? Evolutionism and creationism in Britain and the U.S.* Paper presented at the American Anthropological Association, 95th Annual Meeting, San Francisco, CA.

Collins, H. M., & Pinch, T. (1979). The construction of the paranormal: Nothing unscientific is happening. In R. Wallis (Ed.), *On the margins of science: The social construction of rejected knowledge* (pp. 237–269). Keele, England: University of Keele Press.

Collins, H. M., & Pinch, T. (1993). *The Golem: What everyone should know about science.* Cambridge, England: Cambridge University Press.

Cooper, B. (1995). *After the Flood: The early post-Flood history of Europe.* Chichester, England: New Wine Press.

Cracraft, J. (1983). The scientific response to creationism. In M. C. LaFollette (Ed.), *Creationism, science and the law: The Arkansas case* (pp. 138–147). Cambridge, MA: MIT Press.

Crook, S., Pakulski, J., & Waters, M. (1992). *Postmodernization: Change in advanced society.* London: Sage.

Darwin, C. (1859). *The origin of species by means of natural selection: Or the preservation of favoured races in the struggle for life.* London: Odhams Press.

Davie, G. (1994). *Religion in Britain since 1945: Believing without belonging.* Oxford, England: Blackwell.

Davie, G. (1995). Competing fundamentalisms. *Sociology Review, 4*, (4) 2–7.

DeCamp, L. S. (1968). *The great monkey trial.* New York: Doubleday.

Docherty, T. (Ed.). (1993). *Postmodernism: A reader.* Hemel Hempstead, England: Harvester Wheatsheaf.

Dolby, R. G. A. (1987). Science and pseudo-science: The case of creationism. *Zygon, 22*, 195–212.

Dolby, R. G. A. (1996). *Uncertain knowledge: An image of science for a changing world.* Cambridge, England: Cambridge University Press.

Dolphin, W. D. (1996). A brief critical analysis of scientific creationism. In D. B. Wilson (Ed.), *Did the Devil make Darwin do it? Modern perspectives on the creation–evolution controversy* (2nd ed., pp. 19–36). Ames, IA: Iowa State University Press.

Drew, P., & Heritage, J. (1992). *Talk at work: Interaction in institutional settings.* Cambridge, England: Cambridge University Press.

Dunbar, R. (1995). *The trouble with science.* London: Faber.

Durant, J. (1990). Copernicus and Conan Doyle; or, why should we care about the public understanding of science. *Science & Public Affairs, 5,* 7–22.

Durant, J., Evans, G., & Thomas, G. (1989, July 6). The public understanding of science. *Nature, 340,* 11–14.

Durant, J., Evans, G., & Thomas, G. (1992). Public understanding of science in Britain: The role of medicine in the popular representation of science. *Public Understanding of Science, 1,* 161–182.

Durkheim, E. (1964). The dualism of human nature and its social conditions. In K. H. Wolff (Ed.), *Emile Durkheim et al.: Essays on sociology and philosophy* (pp. 325–340). New York: Harper & Row.

Durkheim, E. (1976). *The elementary forms of the religious life.* (J. W. Swain, Trans., 2nd ed). London: Allen & Unwin.

Eagleton, T. (1983). *Literary theory: An introduction.* Oxford, England: Blackwell.

Edley, N. (1993). Prince Charles—Our flexible friend: Accounting for variations in constructions of identity. *Text, 13,* 397–422.

Edwards, D., Ashmore, M., & Potter, J. (1995). Death and furniture: The rhetoric, politics and theology of bottom line arguments against relativism. *History of the Human Sciences, 8,* 25–49.

Edwards, D., & Potter, J. (1992). *Discursive Psychology.* London: Sage.

Evans, G., & Durant, J. (1989). Understanding of science in Britain and the USA. In R. Jowell, S. Witherspoon, & L. Brook (Eds.), *British social attitudes: Special international report* (pp. 105–119). Aldershot, England: Gower.

Fairclough, N. (1992). *Discourse and social change.* Cambridge, England: Polity.

Fish, S. (1980). *Is there a text in this class? The authority of interpretive communities.* Cambridge, MA: Harvard University Press.

Forrester, M. (1996). *Psychology of language: A critical introduction.* London: Sage.

Foster, H. (Ed.). (1985). *Postmodern culture.* London: Pluto.

Fuller, S. (1997). *Science.* Buckingham, England: Open University Press.

Garfinkel, H. (1967). *Studies in ethnomethodology.* Englewood Cliffs, NJ: Prentice-Hall.

Geering, E. E. M., & Turner, C. E. A. (1989). In the beginning. (Pamphlet 263). Portsmouth, England: Creation Science Movement.

Gellner, E. (1964). *Thought and change.* Chicago: University of Chicago Press.

Gellner, E. (1974). *Legitimation of belief.* Cambridge, England: Cambridge University Press.

Gellner, E. (1992a). *Postmodernism, reason and religion.* London: Routledge.

Gellner, E. (1992b). *Reason & culture: The historic role of rationality and rationalism.* Oxford, England: Blackwell.

Gerlovich, J. A., & Weinberg, S. L. (1996). The battle in Iowa: qualified success. In D. B. Wilson (Ed.), *Did the Devil make Darwin do it? Modern perspectives on the creation–evolution controversy* (2nd ed., pp. 189–205). Ames, IA: Iowa State University Press.

Giddens, A. (1993). *New rules of sociological method: A positive critique of interpretative sociologies* (2nd ed.). Cambridge, England: Polity.

Gieryn, T. F. (1983). Boundary-work and the demarcation of science from non-science: Strains and interests in professional ideologies of science. *American Sociological Review, 48,* 781–795.

Gieryn, T. F., Bevins, G. M., & Zehr, S. C. (1985). Professionalization of American scientists: Public science in the creation/evolution trials. *American Sociological Review, 50,* 392–409.

Gilbert, G. N., & Mulkay, M. (1983). In search of the action. In G. N. Gilbert & P. Abell (Eds.), *Accounts and action: Surrey conferences on sociological theory and method 1* (pp. 8–34). Aldershot: Gower.

Gilbert, G. N., & Mulkay, M. (1984). *Opening Pandora's box: A sociological analysis of scientists' discourse.* Cambridge, England: Cambridge University Press.

Gilkey, L. (1987). Religion and science in an advanced scientific culture. *Zygon, 22,* 165–178.

Goatly, A. (1997). *The language of metaphors.* London: Routledge.

Godfrey, L. R. (Ed.). (1983). *Scientists confront creationism.* London: W. W. Norton & Co.

Gray, C. H. (1995). *The cyborg handbook*. New York: Routledge.

Green, R. W. (Ed.). (1985). Media sensationalism and science: The case of the criminal chromosome. In T. Shinn & R. Whitley (Eds.), *Expository science: Forms and functions of popularisation* (pp. 139–161). Dordrecht: D. Reidel.

Grint, K., & Woolgar, S. (1997). *The machine at work: Technology, work and organization*. Cambridge, England: Polity.

Gross, A. G. (1994). The roles of rhetoric in the public understanding of science. *Public Understanding of Science, 3,* 3–23.

Habermas, J. (1971). Technology and science as "ideology". (J. Shapiro, Trans.). In *Toward a rational society: Student protest, science and politics* (pp. 81–127). London: Heinemann.

Habermas, J. (1973). What does a crisis mean today? Legitimation problems in late capitalism. *Social Research, 40,* 643–667.

Habermas, J. (1984). *The theory of communicative action. Vol. 1: Reason and the rationalization of society.* (T. McCarthy, Trans.). London: Heinemann.

Habermas, J. (1987a). *Knowledge & Human Interests* (2nd ed.). (J. Shapiro, Trans.). Cambridge and Oxford, England: Polity & Blackwell.

Habermas, J. (1987b). *The theory of communicative action. Vol. 2: Lifeworld and system, a critique of functionalist reason.* (T. McCarthy, Trans.). Cambridge, England: Polity.

Halfpenny, P. (1988). Talking of talking, writing of writing: Some reflections on Gilbert and Mulkay's discourse analysis. *Social Studies of Science, 18,* 169–182.

Halfpenny, P. (1989). Reply to Potter and McKinlay. *Social Studies of Science, 19,* 145–152.

Hall, S. (1980). Encoding/Decoding. In S. Hall, D. Hobson, A. Lowe, & P. Willis (Eds.), *Culture, media, language: Working papers in cultural studies, 1972–79.* London: Hutchinson in association with the Centre for Contemporary Cultural Studies, University of Birmingham.

Hamilton, M. (1995). *The sociology of religion: Theoretical and comparative perspectives.* London: Routledge.

Handlin, O. (1965). Science and technology in popular culture. In G. Holton (Ed.), *Science and culture: A study of cohesive and disjunctive forces* (pp. 156–170). Boston: Beacon Press.

Harvey, D. (1990). *The condition of postmodernity: An inquiry into the origins of cultural change.* Oxford, England: Basil Blackwell.

Hawking, S. W. (1988). *A brief history of time: From the Big Bang to black holes.* London: Bantam.

Heelas, P. (1996). *The New Age movement: The celebration of the self and the sacralization of modernity.* Oxford, England: Blackwell.

Heritage, J. (1984). *Garfinkel and ethnomethodology.* Cambridge, England: Polity.

Hess, D. J. (1993). *Science in the New Age: The paranormal, its defenders and debunkers, and American culture.* Madison, WI: University of Wisconsin Press.

Hilgartner, S. (1992). The diet–cancer debate. In D. Nelkin (Ed.), *Controversy: Politics of technical decisions* (2nd ed., pp. 115–129). London: Sage.

Hodge, M. J. S. (1988). England. In T. F. Glick (Ed.), *The comparative reception of Darwinism* (pp. 3–31). Chicago: University of Chicago Press.

Hodge, R., & Kress, G. (1993). *Language as ideology* (2nd ed.). London: Routledge.

Hollis, M., & Lukes, S. (1982). *Rationality and relativism.* Oxford, England: Basil Blackwell.

Holton, G. (1992). How to think about the "anti-science" phenomenon. *Public Understanding of Science, 1,* 103–128.

Holton, G. (1993). *Science and anti-science.* Cambridge, MA: Harvard University Press.

Hornig, S. (1993). Reading risk: Public response to print media accounts of technological risk. *Public Understanding of Science, 2,* 95–109.

Hutchby, I. (1996). *Confrontation talk: Arguments, asymmetries, and power on talk radio.* Mahwah, NJ: Lawrence Erlbaum Associates.

Irwin, A. (1995). *Citizen science: A study of people, expertise and sustainable development.* London: Routledge.

Irwin, A., & Wynne, B. (1996a). Introduction. In A. Irwin & B. Wynne (Eds.), *Misunderstanding science? The public reconstruction of science and technology* (pp. 1–17). Cambridge, England: Cambridge University Press.

Irwin, A., & Wynne, B. (Eds.). (1996b). *Misunderstanding science? The public reconstruction of science and technology.* Cambridge, England: Cambridge University Press.

Jackson, S., & Moores, S. (Eds.). (1995). *The politics of domestic consumption: Critical readings.* Hemel Hempstead, England: Prentice-Hall/Harvester Wheatsheaf.

Jasanoff, S., Markle, G. E., Petersen, J. C., & Pinch, T. (Eds.). (1995). *Handbook of science and technology studies.* London: Sage.

Jefferson, G. (1991). List construction as a task and resource. In G. Psathas & R. Frankel (Eds.), *Interactional competence* (pp. 63–92). Hillsdale, NJ: Lawrence Erlbaum Associates.

Jenkins, H. (1992). *Textual poachers: Television fans and participatory culture.* New York: Routledge.

Katz, I. (1996, April 11). Monkey retrial. *The Guardian,* 2–3.

Kepel, G. (1994). *The revenge of God: The resurgence of Islam, Christianity and Judaism in the modern world.* Trans. A. Braley. Cambridge, England: Polity.

Kitcher, P. (1982). *Abusing science: The case against creationism.* Cambridge, MA: MIT Press.

Knorr-Cetina, K. D., & Mulkay, M. (Eds.). (1983). *Science observed: Perspectives on the social study of science.* London: Sage.

Koestler, A. (1964). *The sleepwalkers: A history of man's changing vision of the universe.* Harmondsworth, England: Penguin.

Kuhn, T. S. (1970). *The structure of scientific revolutions* (2nd ed., enlarged.). Chicago: University of Chicago Press.

Kumar, K. (1995). *From post-industrial to post-modern society: New theories of the contemporary world.* Oxford, England: Blackwell.

LaFollette, M. C. (Ed.). (1983). *Creationism, science and the law: the Arkansas case.* Cambridge, MA: MIT Press.

LaFollette, M. C. (1990). *Making science our own: Public images of science 1910–1955.* Chicago: University of Chicago Press.

Larrain, J. (1979) *The concept of ideology.* London: Hutchinson.

Lash, S. (1990). *Sociology of postmodernism.* London: Routledge.

Lash, S., & Urry, J. (1987). *The end of organised capitalism.* Cambridge, England: Polity.

Latour, B. (1987). *Science in action: How to follow scientists and engineers through society.* Cambridge, MA: Harvard University Press.

Latour, B., & Woolgar, S. (1986). *Laboratory life: The construction of scientific facts* (2nd ed.). Princeton, NJ: Princeton University Press.

Law, J. (1991). Introduction: Monsters, machines and sociotechnical relations. In J. Law (Ed.), *A sociology of monsters: Essays on power, technology and domination* (pp. 1–23). London: Routledge.

Layton, R. (1996, November). *The politics of indigenous "creationism" in Australia.* Paper presented at the American Anthropological Association, 95th Annual Meeting, San Francisco, CA.

Lessl, T. M. (1988). Heresy, orthodoxy, and the politics of science. *Quarterly Journal of Speech, 74,* 18–34.

Lewenstein, B. V. (1995). Science and the media. In S. Jasanoff, G. E. Markle, J. C. Petersen & T. Pinch, (Eds.), *Handbook of science and technology studies* (pp. 343–360). London: Sage.

Livingstone, D. N. (1987). *Darwin's forgotten defenders: The encounter between evangelical theology and evolutionary thought.* Grand Rapids, MI and Edinburgh: William B. Eerdman and Scottish Academic Press.

Locke, S. (1994a). Rationalisation or resource? A study of the use of scientists' discourse by the Creation Science Movement in Britain. Doctoral dissertation, University of Leicester.

Locke, S. (1994b). The use of scientific discourse by creation scientists: Some preliminary findings. *Public Understanding of Science, 3,* 403–424.

Locke, S. (1996a). *The dilemmatic of science in modern culture and the conceit of intellectuals.* In E. Višňovský & G. Gianchi (Eds.), Discourse—Intellectuals—Social communication (pp. 188–211). Bratislava, Slovakia: VEDA.

Locke, S. (1996, November). *Discursive syncretism and the management of political–legal argumentation in national contexts.* Paper presented at the American Anthropological Association, 95th Annual Meeting, San Francisco, CA.

Lynch, M. (1993). *Scientific practice and ordinary action: Ethnomethodology and social studies of science.* Cambridge, England: Cambridge University Press.

Lyon, D. (1994). *Postmodernity.* Buckingham, England: Open University Press.

Lyotard, J.-F. (1984). *The postmodern condition: A report on knowledge.* (G. Bennington & B. Massumi Trans.). Manchester, England: Manchester University Press.

Mackay, H. (1995). Theorising the IT/society relationship. In N. Heap, R. Thomas, G. Einon, R. Mason, & H. Mackay (Eds.), *Information technology and society: A reader* (pp. 41–53). London: Sage.

Macnaghten, P. (1993). Discourses of nature: Argumentation and power. In E. Burman & I. Parker (Eds.), *Discourse analytic research: Repertoires and readings of texts in action* (pp. 52–72). London: Routledge.

Marsden, G. M. (1977). Fundamentalism as an American phenomenon: A comparison with English evangelicalism. *Church History, 46*, 215–232.

Marshall, H., & Raabe, B. (1993). Political discourse: Talking about nationalization and privatization. In E. Burman & I. Parker (Eds.), *Discourse analytic research: Repertoires and readings of texts in action* (pp. 35–51). London: Routledge.

Marx, K., & Engels, F. (1964). *The German ideology.* (S. Ryazanskaya, Trans.). London: Lawrence & Wishart.

McKinlay, A., & Potter, J. (1987). Model discourse: Interpretative repertoires in scientists' conference talk. *Social Studies of Science, 17*, 443–463.

McPhail, M. L. (1996). *Zen in the art of rhetoric: An inquiry into coherence.* Albany, NY: State University of New York Press.

Merton, R. K. (1968a). Science and democratic social structure. In *Social theory and social structure* (2nd ed., pp. 604–615). New York: The Free Press.

Merton, R. K. (1968b). Science and the social order. In *Social theory and social structure* (2nd ed., pp. 591–603). New York: The Free Press.

Merton, R. K. (1970). *Science, technology and society in 17th century England.* New York: Fertig.

Michael, M. (1992). Lay discourses of science: Science-in-general, science-in-particular, and self. *Science, Technology & Human Values, 17*, 313–333.

Michael, M. (1996a). *Constructing identities: The social, the nonhuman and change.* London: Sage.

Michael, M. (1996b). Ignoring science: Discourses of ignorance in the public understanding of science. In A. Irwin & B. Wynne (Eds.), *Misunderstanding science? The public reconstruction of science and technology* (pp. 107–125). Cambridge, England: Cambridge University Press.

Miller, J. D. (1983). Scientific literacy: A conceptual and empirical review. *Daedalus, 112*, 29–48.

Miller, J. D. (1992). Toward a scientific understanding of the public understanding of science and technology. *Public Understanding of Science, 1*, 23–26.

Miller, K. R. (1984). Scientific creationism versus evolution: The mislabelled debate. In A. Montagu, (Ed.), *Science and creationism* (pp. 18–63). Oxford, England: Oxford University Press.

Montagu, A. (Ed.). (1984). *Science and creationism.* Oxford, England: Oxford University Press.

Morley, D. (1974). Reconceptualising the media audience: Towards an ethnography of audiences. (Mimeograph). Birmingham: Centre for Contemporary Cultural Studies, University of Birmingham.

Morley, D. (1980). *The "Nationwide" audience: Structure and decoding.* London: British Film Institute.

Morley, D. (1986). *Family television: cultural power and domestic leisure.* London: Comedia.

Morris, H. M. (Ed.). (1974). *Scientific creationism* (Public School edition). San Diego, CA: CLP Publishers.

Mulkay, M. (1979). *Science and the sociology of knowledge.* London: Allen & Unwin.

Mulkay, M. (1981). Action and belief or scientific discourse? A possible way of ending intellectual vassalage in social studies of science. *Philosophy of the Social Sciences, 11*, 163–171.

Mulkay, M. (1985). *The word and the world: Explorations in the form of sociological analysis.* London: Allen & Unwin.

Mulkay, M., & Gilbert, G. N. (1982). Accounting for error: How scientists construct their social world when they account for correct and incorrect belief. *Sociology, 16*, 165–183.

Mulkay, M., Potter, J., & Yearley, S. (1983). Why an analysis of scientific discourse is needed. In K. D. Knorr-Cetina & M. Mulkay (Eds.), *Science observed: Perspectives on the social study of science* (pp. 171–203). London: Sage.

Myers, G. (1990). *Writing biology: Texts in the social construction of scientific knowledge.* Madison, WI: University of Wisconsin Press.

Nelkin, D. (1982). *The creation controversy: Science or scripture in the schools.* London: W. W. Norton & Co.

Nelkin, D. (1992a). The creation-evolution controversy. In D. Nelkin (Ed.), *Controversy: Politics of technical decisions* (3rd ed., pp. 179–196). London: Sage.

Nelkin, D. (1992b). Science, technology and political conflict: Analyzing the issues. In D. Nelkin (Ed.), *Controversy: Politics of technical decisions* (3rd ed., pp. ix–xxv). London: Sage.

Nettleton, S. (1995). *The sociology of health and illness.* Cambridge, England: Polity.

Numbers, R. L. (1982). Creationism in 20th-century America. *Science, 218,* 538–544.

Numbers, R. L. (1987). The creationists. *Zygon, 22,* 133–165.

Overton, W. R. (1982). Memorandum opinion. In D. Nelkin, *The creation controversy: Science or scripture in the schools* (pp. 201–228). London: W. W. Norton & Co.

Paine, R. (1992). "Chernobyl" reaches Norway: The accident, science, and the threat to cultural knowledge. *Public Understanding of Science, 1,* 261–280.

Parsons, T. (1951). *The social system.* London: Routledge & Kegan Paul.

Parsons, T. (1966). *Societies: Evolutionary and comparative perspectives.* Englewood Cliffs, NJ: Prentice-Hall.

Parsons, T. (1968). *The structure of social action: A study in social theory with special reference to a group of European writers. Volume II.* New York: The Free Press.

Patterson, J. W. (1983). Thermodynamics and evolution. In L. R. Godfrey (Ed.), *Scientists confront creationism* (pp. 99–116). London: W. W. Norton & Co.

Pollner, M. (1974). Mundane reasoning. *Philosophy of the Social Sciences, 4,* 35–54.

Pollner, M. (1987). *Mundane reason: Reality in everyday and sociological discourse.* Cambridge, England: Cambridge University Press.

Pomerantz, A. (1986). Extreme case formulations: A way of legitimizing claims. *Human Studies, 9,* 219–229.

Popper, K. (1952). *The open society and its enemies, Vol. 2: The high tide of prophecy, Hegel, Marx and the aftermath* (2nd ed.). London: Routledge & Kegan Paul.

Popper, K. (1957). *The poverty of historicism.* London: Routledge & Kegan Paul.

Popper, K. (1972). *Objective Knowledge.* Oxford: Clarendon Press.

Popper, K. (1974a). Darwinism as a metaphysical research programme. In P. A. Schilpp (Ed.), *The philosophy of Karl Popper* (pp. 133–143). LaSalle, IL: Open Court.

Popper, K. (1974b). The problem of demarcation. In P. A. Schilpp (Ed.), *The philosophy of Karl Popper* (pp. 976–1048). LaSalle, IL: Open Court.

Potter, J. (1984). Testability, flexibility: Kuhnian values in scientists' discourse concerning theory choice. *Philosophy of the Social Sciences, 14,* 303–330.

Potter, J. (1987). Reading repertoires: A preliminary study of some techniques that scientists use to construct readings. *Science & Technology Studies, 5,* 112–121.

Potter, J. (1988). Cutting cakes: A study of psychologists' social categorizations. *Philosophical Psychology, 1,* 17–33.

Potter, J. (1996). *Representing reality: Discourse, rhetoric and social construction.* London: Sage.

Potter, J., & McKinlay, A. (1989). Discourse–philosophy–reflexivity: comment on Halfpenny. *Social Studies of Science, 19,* 137–145.

Potter, J., & Wetherell, M. (1987). *Discourse and social psychology.* London: Sage.

Potter, J., & Wetherell, M. (1989). Fragmented ideologies: Accounts of educational failure and positive discrimination. *Text, 9,* 175–190.

Potter, J., & Wetherell, M. (1994). Analyzing discourse. In A. Bryman & R. G. Burgess (Eds.), *Analyzing qualitative data.* (pp. 47–66). London: Routledge.

Potter, J., Wetherell, M., & Chitty, A. (1991). Quantification rhetoric—Cancer on television. *Discourse & Society, 2,* 333–365.

Prelli, L. J. (1989a). *A rhetoric of science: Inventing scientific discourse.* Columbia, SC: University of South Carolina Press.

Prelli, L. J. (1989b). The rhetorical construction of scientific ethos. In H. W. Simons (Ed.), *Rhetoric in the human sciences* (pp. 48–68). London: Sage.

Price, D., de Solla (1984). *Little science, big science* (2nd ed.). New York: Columbia University Press.

Psathas, G. (1995). *Conversation analysis: The study of talk-in-interaction.* London: Sage.

Restivo, S. P. (1978). Parallels and paradoxes in modern physics and Eastern mysticism I. *Social Studies of Science, 8,* 143–181.

Restivo, S. P. (1982). Parallels and paradoxes in modern physics and Eastern mysticism II. *Social Studies of Science, 12,* 37–71.

Rose, M. A. (1991). *The post-modern & the post-industrial: A critical analysis.* Cambridge, England: Cambridge University Press.

Rosenberg, C. E. (1961). *No other gods: On science and American social thought.* Baltimore: Johns Hopkins University Press.

Rosenberg, C. E. (1972). Scientific theories and social thought. In B. Barnes (Ed.), *Sociology of science: Selected readings* (pp. 292–305). Harmondsworth, England: Penguin.

Rosevear, D. (1991). *Creation science: Confirming that the Bible is right.* Chichester, England: New Wine Press.

Ross, A. (Ed.). (1996). *Science wars.* Durham, NC: Duke University Press.

Rothman, H., Glasner, P., & Adams, C. (1996). Proteins, plants, and currents: Rediscovering science in Britain. In A. Irwin & B. Wynne (Eds.), *Misunderstanding science? The public reconstruction of science and technology* (pp. 148–157). Cambridge, England: Cambridge University Press.

Royal Society. (1985). *The public understanding of science.* London: Royal Society.

Ruse, M. (1982). *Darwinism defended: A guide to the evolution controversies.* Reading, MA: Addison-Wesley.

Ruse, M. (1989). *The Darwinian paradigm: Essays on its history, philosophy, and religious implications.* London: Routledge.

Sacks, H. (1995). *Lectures on conversation: Volumes I & II.* Edited by G. Jefferson. Oxford, England: Blackwell.

Sacks, H., Schegloff, E., & Jefferson, G. (1974). A simplest systematics for the organization of turn-taking for conversation. *Language, 50,* 696–735.

Schiffrin, D. (1994). *Approaches to discourse.* Oxford, England: Blackwell.

Schneider, D. (1995, July). Darwin denied. *Scientific American, 273,* (1) 8–10.

Schutz, A. (1962). *Collected papers I: The problem of social reality.* The Hague, Netherlands: Martinus Nijhoff.

Shapiro, R. (1986). *Origins: A skeptic's guide to the creation of life on earth.* Harmondsworth, England: Penguin.

Sharma, U. (1992). *Complementary medicine today: Practitioners and patients.* London: Tavistock/Routledge.

Sharrock, W. W., & Anderson, D. C. (1981). Language, thought and reality, again. *Sociology, 15,* 287–293.

Shermer, M. B. (1991). Science defended, science defined: The Louisiana creationism case. *Science, Technology & Human Values, 16,* 517–539.

Shotter, J., & Gergen, K. (Eds.). (1989). *Texts of identity.* London: Sage.

Silverstone, R., & Hirsch, E. (Eds.). (1992). *Consuming technologies: Media and information in domestic spheres.* London: Routledge.

Simons, H. W. (Ed.). (1989). *Rhetoric in the human sciences.* London: Sage.

Smart, B. (1992). *Modern conditions, postmodern controversies.* London: Routledge.

Smith, D. (1978). "K is mentally ill": The anatomy of a factual account. *Sociology, 12,* 23–53.

Snow, C. P. (1964). *The two cultures and a second look* (2nd ed.). Cambridge, England: Cambridge University Press.

Stevenson, N. (1995). *Understanding media cultures: Social theory and mass communications.* London: Sage.

Storer, N. W. (1966). *The social system of science.* New York: Holt, Rinehart & Winston.

Strinati, D. (1995). *An introduction to theories of popular culture.* London: Routledge.

Taylor, C. A. (1992). Of audience, expertise and authority: The evolving creationism debate. *Quarterly Journal of Speech, 78,* 277–295.

Taylor, C. A. (1996). *Defining science: A rhetoric of demarcation.* Madison, WI: University of Wisconsin Press.

Taylor, C. A., & Condit, C. M. (1988). Objectivity and elites: A creation science trial. *Critical Studies in Mass Communications, 5,* 293–312.

Thwaites, T., Davis, L., & Mules, W. (1994). *Tools for cultural studies: An introduction.* South Melbourne, Australia: Macmillan.

Toumey, C. P. (1991). Modern creationism and scientific authority. *Social Studies of Science, 21,* 681–699.

Toumey, C. P. (1994). *God's own scientists: Creationists in a secular world.* New Brunswick, NJ: Rutgers University Press.

Trefil, J., & Hazen, R. M. (1995). *The sciences: An integrated approach.* New York: Wiley.

Tulloch, J., & Jenkins, H. (1995). *Science fiction audiences: Watching Doctor Who and Star Trek.* London: Routledge.

Turkle, S. (1984). *The second self: computers and the human spirit.* London: Granada.

Turner, B. S. (1983). *The body and society: Explorations in social theory.* Oxford, England: Basil Blackwell.

Turner, C. E. A. (1982). A jubilee of witness for creation against evolution by CSM/EPM, 1932–1982. Pamphlet 232. Portsmouth, England: Creation Science Movement.

Turner, R. (1974). *Ethnomethodology: Selected readings.* Harmondsworth, England: Penguin.

van Dijk, T. (Ed.). (1985). *Handbook of discourse analysis* (4 vols.). London: Academic Press.

Volosinov, V. N. (1973). *Marxism and the philosophy of language.* (L. Matejka & I. R. Titunik, Trans.). Cambridge, MA: Harvard University Press.

Wallis, R. (Ed.). (1979). *On the margins of science: The social construction of rejected knowledge.* Keele, England: University of Keele.

Weber, M. (1948). Science as a vocation. In H. H. Gerth & C. W. Mills (Eds. And Trans.), *From Max Weber: Essays in sociology* (pp. 129–156). London: Routledge & Kegan Paul.

Weber, M. (1976). *The Protestant ethic and the spirit of capitalism.* London: Allen & Unwin.

Webster, A. (1991). *Science, technology and society: New directions.* London: Macmillan.

Wetherell, M., & Potter, J. (1988). Discourse analysis and the identification of interpretative repertoires. In C. Antaki (Ed.) *Analysing everyday explanation* (pp. 168–183). London: Sage.

Wetherell, M., & Potter, J. (1992). *Mapping the language of racism: Discourse and the legitimation of exploitation.* London: Harvester Wheatsheaf.

Whalen, M. R., & Zimmerman, D. H. (1990). Describing trouble: Practical epistemology in citizen calls to the police. *Language in Society, 19,* 465–492.

Whitcomb, J. C., Jr., & Morris, H. M. (1969). *The Genesis Flood: The biblical record and its scientific implications.* London: Evangelical Press.

Whitley, R. D. (1983). From the sociology of scientific communities to the study of scientists' negotiations and beyond. *Social Science Information, 22,* 681–720.

Whitley, R. D. (1985). Knowledge producers and knowledge acquirers: Popularisation as a relation between scientific fields and their publics. In T. Shinn & R. Whitley (Eds.), *Expository science: Forms and functions of popularisation* (pp. 3–28). Dordrecht, Netherlands: D. Reidel.

Williams, G., & Popay, J. (1994). Lay knowledge and the privilege of experience. In J. Gabe, D. Kelleher, & G. Williams (Eds.), *Challenging medicine* (pp. 118–139). London: Routledge.

Wilson, B. (1966). *Religion in a secular society: A sociological comment.* London: C. A. Watts.

Wilson, B. (Ed.). (1970). *Rationality.* Oxford, England: Basil Blackwell.

Wilson, D. B. (Ed.). (1996). *Did the Devil make Darwin do it? Modern perspectives on the creation–evolution controversy* (2nd ed.). Ames, IA: Iowa State University Press.

Woodlief, A. M. (1981). Science in popular culture. In M. Thomas Inge (Ed.), *Handbook of American popular culture Vol. 3* (pp. 429–458). Westport, CT: Greenwood.

Wooffitt, R. (1992). *Telling tales of the unexpected: The organization of factual discourse.* Hemel Hempstead, England: Harvester Wheatsheaf.

Woolgar, S. (1983). Irony in the social study of science. In K. Knorr-Cetina & M. Mulkay (Eds.), *Science observed: Perspectives on the social study of science* (pp. 237–266). London: Sage.

Woolgar, S. (1988). *Science: The very idea.* Chichester and London: Ellis Horwood & Tavistock.

Woolgar, S. (1996, September). *Is there a future for the sociology of scientific knowledge?* Paper presented at the "Is there a future for the sociology of scientific knowledge?" Conference, Tavistock Institute, London, England.

Woolgar, S., & Pawluch, D. (1985). Ontological gerrymandering: The anatomy of social problems explanations. *Social Problems, 32,* 214–227.

Wren-Lewis, J. (1983). The encoding/decoding model: Criticisms and redevelopments for research on decoding. *Media, Culture & Society, 5,* 179–197.

Wynne, B. (1991). Knowledges in context. *Science, Technology & Human Values, 16,* 111–121.

Wynne, B. (1992a). Public understanding of science research: New horizons or hall of mirrors? *Public Understanding of Science, 1,* 37–43.

Wynne, B. (1992b). Misunderstood misunderstanding: Social identities and public uptake of science. *Public Understanding of Science, 1,* 281–304.

Wynne, B. (1993). Public uptake of science: A case for institutional reflexivity. *Public Understanding of Science, 2,* 321–337.

Wynne, B. (1995). Public understanding of science. In S. Jasanoff, G. E. Markle, J. C. Peterson & T. Pinch. (Eds.), *Handbook of science and technology studies* (pp. 361–388). London: Sage.

Yates, F. A. (1972). *The Rosicrucian enlightenment.* London: Routledge & Kegan Paul.

Yearley, S. (1985). Vocabularies of freedom and resentment: a Strawsonian perspective on the nature of argumentation in science and the law. *Social Studies of Science, 15,* 99–126.

Yearley, S. (1991). *The Green case: A sociology of environmental issues, arguments and politics.* London: HarperCollins.

Yearley, S. (1994). Understanding science from the perspective of the sociology of scientific knowledge: An overview. *Public Understanding of Science, 3,* 245–258.

Ziman, J. (1991). Public understanding of science. *Science, Technology & Human Values, 16,* 99–105.

BIBLIOGRAPHY OF CSM PAMPHLETS

The following is a complete list by number of the sample of pamphlets used in this study. Although not all are referred to in the body of the text, I thought it might be of value and interest to readers to have this information. All are published by Creation Science Movement (CSM), Portsmouth, England.

11. Watts, N. (1984). Why I believe in creation (abridged).
14. Dewar, D. (nd.). Do vestigial organs exist?
43. Johannsen, E. (1980). The phenomena of nest-making inexplicable by the evolution–mechanism (new ed.).
58. Turner, C. E. A. (1986). Scientific method and evolution theory (abridged).
62. Betts, E. H. (1958). Entropy disproves evolution (revised).
74. Turner, C. E. A. (1973). Horse evolution (new ed.).
76. Turner, C. E. A. (1973). Archaeopteryx, a bird: No link (new ed.).
80. Tilney, A. G. (1980). Darwin and Christianity (new ed.).
89. Tier, W. H. (1971). Creation or evolution.
90. Acworth, B., (1990). Evolution: The mammoth and the Flood (abridged).
99. Cousins, F. W. (nd.). A note on the diatomaceae and an inability to reconcile them with evolution.
100. Carron, T. W. (1973). Partnership: Planned or accidental? (new ed.).
104. Turner, W. H. (1963). Evolution and miracles.
111. Howitt, J. R. (1980). Karl Marx as an evolutionist (new ed.).
112. Whitcomb, J. C., Jr. (1964). God's Double Revelation in scripture and science? Biblical inerrancy and the Double Revelation Theory (with special reference to the origin of the solar system) (abridged).
114. Cousins, F. W. (1965). Cetacea.
131. Tilney, A. G., & Tilney, D. (1965). The skin.
141. Howe, G. F. (1966). Fossil plants no evidence of evolution.
151. Bowden, M. (1978). Fictional reconstructions of man's (supposed) ancestors (revised).
160. Shelley, J. E. (nd.). The Flood: A short vindication of its historicity, universality and present significance.
171. McM., N. (1969). Ophidia: The problems snakes present to the evolutionists.
177. Turner, C. E. A. (1970). Teleology: Purpose everywhere.
183. Tilney, A. G. (1970). Did Darwin disprove Genesis?
185. Carron, T. W. (1971). The breeding instinct not evolved.
186. Shute, E. (1969). Evolution in the glare of new knowledge.
190. Nicholls, J. (1971). Bacterium E. Coli v. evolution.
195. Turner, C. E. A. (1972). Trace elements in the creation.
197. Turner, C. E. A. (nd.). Dinosaurs: Without descent or descendants.
201. Morrell, R. W. (nd.). Evolutionary contradictions and geological facts.
204. Turner, C. E. A. (nd.). An eminent modern biologist criticises evolutionists.

207. Gower, D. B. (1987). Radiometric dating methods.
209. Radcliffe-Smith, A. (nd.). Biology in the Bible.
210. Gish, D. T. (1975). Creation, evolution and public education.
211. Radcliffe-Smith, A. (nd.). Pollination.
213. Turner, C. E. A. (nd.). The domestication of animals.
214. Rosevear, D. T. (nd.). Enzymes: The assembly-line workers within us.
215. Nevins, S. E. (1977). The origin of coal.
216. Radcliffe-Smith, A. (1977). Orchids.
219. Rosevear, D. T. Genealogies and early man.
221. Nevins, S. E. (1979). The ocean says No! to evolution.
222. Turner, C. E. A. (1979). The camel: Created or evolved?
223a. Turner, C. E. A. (1980). An eminent surgeon and evolution.
223b. Wakeley, C. (1980). A surgeon looks at evolution.
224. Rosevear, D. T. (1980). Scientists critical of evolution.
225. Turner, C. E. A. (Ed.). (1981). Space scientists and evolution.
226. Butt, S. M. (1981). Insect flight: Testimony to creation (abridged).
227. Jones, A. (1981). The genetic integrity of the kinds: A working hypothesis (abridged).
228. Bowden, M., & Collyer, J. V. (Eds.). (1982). Quotable quotes for creationists.
229. Pitman, M. (1982). Creation and science.
231. Angus, R. A. G. (1982). Birds: Created, not evolved!
232. Turner, C. E. A. (1982). A jubilee of witness for creation against evolution by CSM/EPM, 1932–1982.
233. Rosevear, D. T. (1987). The bombardier beetle.
234. Cooper, W. R. (1986). Miocene Man (revised).
236. Bowden, M. (1983). The recent change in the tilt of the earth's axis.
237. Ollerenshaw, K. (nd.). Adam's animals: The Genesis kinds.
239. Wilson, C. (1984). Language: Was it evolved?
240. Gray, I. (1984). Historical geology.
241. Howe, G. F. (1985). The biological macroevolutionary origins model is not a "scientific theory."
242. Rendle-Short, J. (1986). Creation: The foundation of the Gospel.
243. Turner, C. E. A. (1985). Creation in Isaiah.
244. Cooper, B., & Bowden, M. (1985). More on "Miocene Man.".
245. Peachey, F. J. (1986). Creation in Genesis.
246. Wilders, P. (1986). Divine geography.
247. Grantham-Hill, B. W. (1989). The world of bats.
248a. Filmer, W. E. (1986). The palisade moth.
248b. Filmer, W. E. (1986). Evolution and morals (abridged).
249. Rosevear, D. T. (1986). Creation recreated.
250. Sunderland, L. D., and Parker, G. E. (1986). Evolution? Prominent scientist reconsiders.
251. Chapman, G. (1987). Our unique planet.
252. Austin, S. A. (1990). Mount St. Helen's and catastrophism (revised).
253. Famularo, S. (1986). Food for thought.
254. Price, R. (1986). Science and the Bible.
255. Filmer, W. (1987). Feathers—Wonders of creation.
256. Bowden, M. (1987). The speed of light–The second monograph.
257. Grantham-Hill, B. W. (1988). Metamorphosis: The world of butterflies.
258. Wright, V. (1988). Evolution—Some rheumatological riddles.
259. Gish, D. T. (1988). Is it possible to be a Christian and an evolutionist?
260. Watson, D. C. C. (1988). Dare we reinterpret Genesis?
261. Rosevear, D. T. (1988). Molecular biology disproves evolution.
262. Bowden, M. (1988). Decrease in the speed of light.
263. Geering, E. E. M., & Turner, C. E. A. (1989). In the beginning.
264. Doolan, R. (1989). Honey bees.
265. Chapman, G. (1989). Our young universe.
266. Ham, K. (1989). The Genesis foundation.
267. Croft, L. (1989). The myth of chemical evolution.
268. Grantham-Hill, B. (1989). The world of ants.

269. Cousins, F. W. (1990). *The eye.*
270. Chapman, G. (1989). *Geological fallacies.*
271. Foucher, J. (1990). *Birds: A special creation.*
272. Rosevear, D. T. (1990). *Creation and the laws of science.*
273. Garner, P. (1990). *The moon: Evidence for special creation.*
274. Cooper, B. (1991). *After the Flood: The witness of early European history.*
275. Curtis, D. (1991). *Dinosaurs.*
276. Gitt, W. (1991). *Information: The third fundamental quantity.*
277. Collyer, J. V. (1991). *Evolutionists debunk evolution.*
278. Rosevear, D. T. (1991). *Genetics and creation.*
279. White, A. J. M. (1991). *How old is the earth?*
280. Cooper, B. (1992). *Anglo-Saxon dinosaurs.*
281. Berthault, G. (1992). *The laying down of marine sediments—a revolutionary new perspective.*
282. Grace, P. (1992). *The culture gap.*
283. Rosevear, D. T. (1992). *The Ice Age and the biblical time-scale.*
284. Rosevear, D. T. (1992). *Genesis in Chinese calligraphy.*
285. Watson, D. C. C. (1992). *Exodus miracles and creation.*
286. Knopp, B. B. (1993). *Hurdles of evolution theory.*
287. Geering, E. (1993). *The origin of species.*
288. Fisher, G. A. (1993). *A possible Flood, Ice-Age and earth division mechanism.*
289. Cooper, B. (1993). *Ancient calendars and the age of the earth.*
290. Watson, D. C. C. (1993). *Paley in perspective.*
291. Johannsen, E., & Carron, T. W. (1993). *Instincts and creation* (revised).
292. Grantham-Hill, B. (1994). *The world of woodpeckers.*
293. Gish, D. T. (1994). *Challenge of the fossil record.*
294. Simmonds, G. (1994). *Anthropoid features in the animal world: Characteristics humans share with other creatures.*
295. Aris, A. P. (1994). *Genesis as history.*
296. Watson, D. C. C. (1994). *The end is nigh.*
297. Scheven, J. (1994). *After its kind.*
298. Hamblin, T. (1995). *On whose authority? Contrasting the authority of scientists and of Scripture.*
299. Rosevear, D. T. (1995). *The first seven days of the universe.*
300. Chapman, G. (1995). *Guide to transitional fossils.*
301. Nicholls, P. (1995). *Population growth and the time-span of human history.*
302. Simmonds, G. (1995). *Sharks: More than just jaws.*
303. Bluer, P. (1995). *Numbers in Scripture.*
304. Rahme, F. Abou. (1996). *Evidence for Noah's Ark and the Flood.*
305. Garner, P. (1996). *On the rocks: Evolution and the age of the earth.*
306. Rosevear, D. T. (1996). *Origin of man.*
307. Cooper, B. (1996). *The stooping Rhodesian Man fraud.*
308. Bowden, M. (1996). *The earth, moon and tides.*

Author Index

223

Chitty, A., 61, 108, 125, *215*
Coleman, S. M., 51, 80, *210*
Collins, H. M., 73, 114, 179, *210*
Collyer, J. V., 50, *220, 221*
Condit, C. M., 129, *216*
Condor, S., 9, 20, 39, 202, *210*
Cooper, B., 47, *210, 220, 221*
Cooper, W. R., *220*
Cousins, F. W., 124, *219, 220*
Cracraft, I., 92, *210*
Crook, S., 187, *210*
Croft, L., 134, *220*
Curtis, D., *221*

D

Darwin, C., *210*
Davie, G., 62, 150, *210*
Davis, L., 86, *216*
DeCamp, L. S., 60, *210*
de Solla, 5, *215*
Dewar, D., *219*
Docherty, T., 187, *210*
Dolby, R. G. A., 11, 76, *210, 211*
Dolphin, A. D., 52, *211*
Doolan, R., *220*
Drew, P., 36, *211*
Dunbar, R., 10, 14, 133, 154, *211*
Durant, J., 11, 16, 17, *211*
Durkheim, E., 21, *211*

E

Eagleton. T., 81, *211*
Edge, D., 11, 31, *209, 210*
Edley, N., 41, 81, 124, *211*
Edwards, D., 9, 20, 33, 35, 39, 123, 202, *210, 211*
Engels, F., 21, *214*
Evans, G., 11, 17, *211*

F

Fairclough, N., 37, 38, *211*
Famularo, S., *220*
Filmer, W. E., *220*
Fish, S., 81, *211*
Fisher, G. A., *221*
Forrester, M., 81, 154, *211*
Foster, H., 187, *211*

Foucher, J., *221*
Fuller, S., 16, *211*

G

Gane, M., 9, 20, 39, 202, *210*
Garfinkel, H., 37, 138, *211*
Garner, P., 130, *221*
Geering, E. E. M., 61, 62, *211, 220, 221*
Gellner, E., 24, 25, 26, 175, *211*
Gergen, K., 32, *216*
Gerlovich, J. A., 52, *211*
Giddens, A., 174, *211*
Gieryn, T. F., 7, 12, 34, 54, 71, 73, 78, 107, 124, 126, *211*
Gilbert, G. N., 16, 36, 37, 40, 41, 46, 81, 100, 101, 122, 202, *211, 214*
Gilkey, L., 53, 57, 64, 65, 66, 77, *211*
Gish, D. T., 89, 128, *220, 221*
Gitt, W., *221*
Glasner, P., 34, *216*
Goatly, A., 116, *211*
Godfrey, L. R., 47, *211*
Gower, D. B., *219*
Grace, P., *221*
Grantham-Hill, B., *220, 221*
Gray, C. H., 11, *212*
Gray, I., *220*
Green, R. W., 133, *212*
Grint, K., 11, *212*
Gross, A. G., 6, 11, *212*

H

Habermas, J., 22, 39, 126, 136, 176, 177, 178, 179, 180, 181, 182, 183, 201, *212*
Halfpenny, P., 81, *212*
Hall, S., 81, *212*
Ham, K., 95, 141, *220*
Hamblin, T., *221*
Hamilton, M., 10, *212*
Handlin, O., 18, *212*
Harvey, D., 187, *212*
Hawking, S. W., 164, 165, 167, *212*
Hazen, R. M., 165, *216*
Heelas, P., 174, *212*
Henry, J., 34, *209*
Heritage, J., 36, 138, *209, 211, 212*
Hess, D. J., 11, 40, 204, *212*
Hilgartner, S., 133, *212*
Hirsch, E., 11, *216*

Subject Index

Printed in the United States
by Baker & Taylor Publisher Services

Printed in the United States
by Baker & Taylor Publisher Services